Diabetes:
Alternative Thoughts

Diabetes:
Alternative Thoughts

Dr. Manoj Kumar

PARTRIDGE
A Penguin Random House Company

To order additional copies of this book, contact
Partridge India
000 800 10062 62
www.partridgepublishing.com/india
orders.india@partridgepublishing.com

CONTENTS

DEDICATION

This book is dedicated

To my parents

My ever-supporting wife

Loving daughter

To millions of people suffering from diabetes

And many millions on the verge of it.

INVOCATION

Aum sarve bhavantu sukhinah,
sarve santu niramayah,
sarve bhadrani pashyantu,
ma kashchit dukha bhagbhavet.

Om, may all become happy,

May all be free from illness,

May all see what is auspicious,

May no one suffer.

PRAYER OF ENLIGHTENMENT

Asato Maa Sad Gamaya,

Tamaso Maa Jyotir Gamaya,

Mrityor Maa Amritam Gamaya,

Om Shanti Shanti Shanti.

Oh Lord,

Lead me from untruth to truth,

From darkness to light,

From death to immortality,

Peace, peace, peace.

DISCLOSURE

The information provided in this book is designed to provide helpful information on the subjects discussed. This book is not meant to be used, nor should it be used, to diagnose or treat any medical condition. For diagnosis or treatment of any medical problem, consult your own physician. The publisher and author are not responsible for any specific health or allergy needs that may require medical supervision and are not liable for any damages or negative consequences from any treatment, action, application, or preparation to any person reading or following the information in this book. References are provided for informational purposes only and do not constitute endorsement of any websites or other sources. Readers should be aware that the websites listed in this book may change.

ACKNOWLEDGEMENTS

It is a privilege to publish my first book on a common topic which touches millions of people around the world. Over the years, many patients used to approach me for advice as they find it difficult to control their blood sugar. Many were looking for a magic herbal pill, which I didn't have. Most of them were surprised when they realised the array of options available but didn't try as they were either ignorant or not confident enough to pursue them. Let me express my gratitude to those people who stood firm on their faith with the integrative approaches.

Special thanks to Dr U. Acharya whom I consider as my mentor, for enlightening and encouraging me to implement dietary advice in my practice. I am sure he is always a role model for many physicians who worked with him.
Thanks to Dr Ryan Penny being an inspiration for writing a book and accepting my request to write the foreword.

When the manuscript was done, I wanted it to be read at least by a couple of people who could suggest corrections and inputs. Sincere gratitude to my friend, Dr Muralikrishnan (Physician, Indian Railways) for accepting the challenge of going through the pages to ensure there are no medical errors. Since this book is not meant to target medical professionals alone, inputs from non-medical people were essential to know if it is readable and can be followed by people without medical back ground. Thanks to my friend in Dubai, Mr Ramesh Venkit, for taking the effort to go through the technological jargon and for giving me his valuable suggestions.

Thanks to Mr Vinayaraj (Bank Officer, Kerala) for agreeing to use his photos in this book.

Finally, I would like to thank Partridge publishers for helping me to bring it out in book format.

FOREWORD

Dr. Manoj is a man I know well. For five years we worked closely together at a clinic in Dubai. Thinking back on that time, I recall three things about him that stand out and also support your decision to read his book.

The first is his *attention to detail*. Manoj is a man who reads, studies, and thinks deeply about complex medical conditions. Why does he do this? To provide his patients with the best information and care possible, that's why. Despite his initial intention of presenting a 'simple' summary of diabetes and its management for laypeople, his devotion to detail has not escaped his book. That's a good thing, because now we have a resource valuable for patient and practitioner alike.

The second thing that stands out about the author is his *sincere concern* for people who entrust their health to him. Manoj genuinely wants to assist diabetes sufferers overcome the obstacles the disease poses to their well-being. This is evident in his dedication right at the beginning of his book, where he includes you among the millions of other people he longs to help.

The third, seemingly paradoxical, thing that stands out about Manoj is his *humility*. Unlike many, who write to be in the limelight, I'm convinced that Manoj has written to avoid it. This book is his way of quietly offering suggestions based on his vast knowledge and experience. He has no desire to push his opinions or force-feed his solutions. Instead, he humbly extends a practical means of tackling a disease that could steal your life. In light of this, you would do well to take my advice—read his.

Ryan Penny
Clinical Homeopath, Nutritionist, and Author

Diabetes is a monster that feeds on a lifestyle promoted by a world in pursuit of instant satisfaction.
(Ryan Penny)

PROLOGUE

At least some of you would be wondering about the motives of writing a book like this, when many similar titles are already in the market. For this I have to go back in time in my medical career.

Soon after finishing medical school, I got an opportunity to participate in a clinical trial for diabetes. Aim of the trial was to find out the efficacy of a herbal compound compared with the standard treatment. It gave me a chance to realise the magnitude of complications for this killer disease. Until then, as a student and intern, I used to come across patients having diabetic complications: ketoacidosis, diabetic gangrene, foot ulcers, or amputation. Those cases never gave me an opportunity to understand the lifestyle and dietary habits of diabetics. It was during my practice in UAE, where 25% of the population is diabetic, I realized the impact of physical activity and healthy eating habits in diabetics. I still continue to advise people for improving their lifestyle and eating habits and try to manage patients with an integrative approach.

Let me make it clear that I am not a diabetologist/ endocrinologist. But during the last several years of my practice both as specialist in Physiatry (Physical Medicine) and qualified Alternative Medicine practitioner (Indian Herbal system/Ayurveda), I managed to spend time in educating myself and my patients about lifestyle changes in diabetes. So I thought it would be unfair if I keep this knowledge to myself without sharing with the public.

Most of the ideas given in this book is neither new nor breakthrough. These are known to many physicians but either they don't believe in them or are too busy to include them in their practice. The information is based on scientific facts and I tried to include available references. The normal convention for treatment in any kind of disorder is to start with Conventional Medicine (aka Modern Medicine) and try alternatives when it is not effective or when the person gets disillusioned with the medications or its complications. An interesting fact which most people are unaware is that the so-called Alternative system of approach is the mainstream treatment in many parts of the world. More than 80% of people use some form or the other of herbal or alternative medicines all over the world. The mere fact that traditional systems of medicine like Ayurveda and Chinese medicine have stood the test of time itself gives credentials to their efficacy and trust among people. We know that there is scarcity of valid scientific studies in alternative medicine. In the recent times, we see a healthy change as more researches are being done in this field.

Herbal medicines are considered as intellectual property of the tribe, country, or system of medicine. Since getting an exclusive patent for herbal compounds is difficult, most of the pharmaceutical companies are not interested doing researches in this field.

In the following pages, I will try to prove the importance of alternative approaches with ample scientific evidence. I am sure that this book would serve as an eye opener for people suffering from diabetes and give them confidence in trying integrative approaches in management.

CHAPTER 1

Introduction

Type 2 diabetes mellitus, otherwise mentioned simply as diabetes is one of the leading killer diseases in the modern world. It is calculated that every two seconds a person is detected with diabetes somewhere in the world, and every four seconds a person is dead because of it.

The figures given by different organisations are startling. As per the latest WHO estimate, around 345 million people worldwide are suffering from this problem, and the numbers are supposed to double in 2020. The disease has got an enormous impact on the world economy.

The increased surge in number of diabetics could be due to various reasons: the swelling of waistline of the so-called middle-income group, increasing number of people adopting 'Western lifestyle', improved blood-testing facilities in more populous and developing countries, and the lowering of cut off point for diabetic detection. In developed and affluent parts of developing countries, the bulging waistline and couch-potato lifestyle are the main culprits along with overconsumption of high refined carbs. It is ironic that plain water is offered rarely in many of the food outlets, and in most of them, water is costlier than soda pops!

A peek into history of medicine would help you to understand that diabetes is not a disease of the modern world but was prevalent among the affluent people of olden times. In recent times, we see a shift in this pattern as more poor people are becoming obese and diabetic than the wealthy and affluent.

'Diabesity' is a new term coined for indicating the coexistence of obesity and diabetes in people. 'Metabolic Syndrome' is the name given when diabetes, high blood pressure, obesity (bulging waistline), and high blood lipids (this is a controversial topic!) coexist in a person.

I would be touching more in detail about these as you progress through the book. This book focuses about the most common one, that is, type 2 diabetes, not the type 1 or insulin-dependent diabetes.

Type-2 diabetes begins predominantly with insulin resistance (1); a condition characterised by the inability of cells to respond to the action of insulin, i.e. transporting glucose from the bloodstream into muscle, fat, and liver cells (2). In order to overcome the state of insulin resistance, the body creates a compensatory increase in insulin secretion from the pancreatic β cells (hyperinsulinemia) by which it thinks to solve the problem. However, in the long run, β-cell mass and function progressively decline (3).

Type 2 diabetes is characterised by combination of peripheral insulin resistance and inadequate insulin secretion by the pancreatic β cells (special type of cells on the tail of pancreas—an organ sitting on the back of our upper belly). Insulin resistance is the stage in which body is not able to utilise the secreted insulin either due to damage of insulin receptors in the tissues (like muscle, liver, kidney, and fat cells) or due to inadequate number of receptors. Enzymes need receptors in the cells for their action, and they work similar to lock and key mechanism. In order for the lock to be opened, you need a proper set of keys. But if the keys are damaged or you use the wrong keys, the lock would not open. Similarly insulin has to attach to the cell receptors so that gates for glucose are opened and it can enter the cell. If the receptors are damaged or if they are less in number, the secreted insulin cannot do its job, so excess levels of glucose is accumulated in the blood stream.

Let me explain this situation with a football ground. Imagine the gates of the ground are having electronic sensors and visitors need to have access cards to enter the stadium. Now consider a situation where there is some malfunction in the scanners to read the access cards, which result in locked gates and people flock outside whereas the stadium remains empty. Likewise, the improper functioning of the insulin receptors result in accumulation of glucose outside the cell than inside, where it is needed most. This is what we call diabetes. In the long run, the excess blood sugar can cause damage to various organs, and complications of diabetes ensue.

The disorder of insulin resistance never happens overnight. It happens over a period of years as a result of changes in the system which leads to insulin resistance. The body starts pumping more insulin thinking that it is not enough to control sugar, which results in pancreatic fatigue and inadequate insulin secretion. It has been found that pancreas is atrophied (shrunken) in people with prolonged Type 1 diabetes and reduced in number and volume in type 2 diabetes due to apoptosis (programmed cell death) (3). In obese individuals with impaired glucose, the β cell mass is 50% less than that of healthy people (4).

Nobody becomes diabetic overnight, as the perpetuating events start many years before detection. Some people are labelled as impaired fasting glucose as their fasting blood sugar values are below 126 mg/dL—the threshold for labelling a person as diabetic (as per WHO and ADA guidelines)—but higher than normal fasting glucose levels (above 110 mg/dL as per WHO guidelines; above 100 mg/dL as per ADA guidelines). The chances of these people heading towards full-blown diabetic in the future is very high, provided they do not listen to the warning signs and adjust their lifestyle and eating habits.

Many a times, I used to wonder that the difference between the so-called diabetic and impaired fasting or the so-called normal and impaired fasting is so negligible yet both practitioners and the patients cling on to these figures. Imagine that one person is having fasting blood sugar of 126 mg/dL and other one having 125 mg/dL, or one having 109 and other having 110? The one who got the label 'normal' would be happy and the other is given the impression that his life is doomed. The inclusion criteria in guidelines are developed on the basis of population studies, and they are not tailor-made for an individual. But the problem arises when these guidelines are implemented without considering geographical, ethnical, and genetic make-up of different populations.

In the following chapters, we try to explore in detail the various factors which cause metabolic imbalances and diabetes.

- Diabetes is a major 'epidemic'.
- Western lifestyle attributed to its surge.
- Insulin is secreted by β cells in pancreas.
- Inadequate insulin or improper utilisation of insulin can result in diabetes.
- Insulin resistance is an important step in diabetes development.
- Nobody becomes diabetic overnight; it is the result of ongoing changes in response to our bad lifestyle and dietary habits.
- There is clustering of diabetes, obesity, high blood cholesterol in the population.
- As per current guidelines, a fasting blood sugar >110 mg/dL is impaired fasting sugar (100 as per ADA).
- Fasting sugar $\geq$126 mg/dL or HbA1c $\geq$ 6.5% is diabetes.
- High risk for HbA1c values 6-6.4%.

References

1. Lillioja S, Mott DM, Spraul M, Ferraro R, Foley JE, Ravussin E, Knowler WC, Bennett PH, Bogardus C. Insulin resistance and insulin secretory dysfunction as precursors of non-insulin-dependent diabetes mellitus. Prospective studies of Pima Indians. N Engl J Med. 1993;329(27):1988-92; PMID: 8247074

2. Festa A, Williams K, D'Agostino Jr. R, Wagenknecht LE, Haffner SM. The natural course of beta-cell function in non-diabetic and diabetic individuals: the Insulin Resistance Atherosclerosis Study. Diabetes. 2006;55(4):1114-20; PMID: 16567536

3. William AJ, Thrower SL, Sequeiros IM, Ward A, Bickerton AS, Triay JM, Callaway MP, Dayan CM. Pancreatic volume is reduced in adult patients with recently diagnosed type 1 diabetes. J Clin Endocrinol Metab. 2012;97(11):E2109-13

4. Butler AE, Janson J, Bonner-Weir S, Ritzel R, Rizza RA, Butler PC. Beta cell deficit and increased beta cell apoptosis in humans with type 2 diabetes. Diabetes. 2003;52(1):102-10

CHAPTER 2

What Causes Diabetes?

It is a difficult question to address.

As I mentioned before, nobody becomes diabetic overnight. It is like germination of a seed. There are plenty of factors that determine the germination and subsequent growth of a seed. Seed factors such as vitality, genotype, and maturity and environmental factors like ambient temperature, moisture, and sunlight are found to be important among them. Germination takes place when these conditions are satisfied. We assume the seed has grown when the bud shows up at the soil top. The appearance of bud on the surface does not reflect the beginning of germination process. This might have started the moment the seed was planted and watered but was not evident from the surface. So we consider the seed has grown when the sprout appears above the soil. Likewise, the pathophysiological changes leading on to a particular disease start years before it is evident or detected. So, diabetes never develops overnight.

The perpetuating events begin when a person with predisposition to diabetes start taking high glycaemic index foods (intake of high glucose releasing foods) on a regular basis without adequate physical activity. The resulting high blood sugar level causes stress in tissues and contributes to high fat deposition (the destiny of most of the sugars is conversion into fat. But for fruit sugars, the pathway is short as they are converted directly into triglycerides). The resultant high sugar stimulates pancreas for extra insulin secretion. When fat is deposited in organs like liver, kidney, muscle, and pancreas (where insulin is utilised), the action of insulin in these tissues is reduced, which leads on to an insulin excess (insulin resistance). Finally, it results in the destruction of pancreatic beta cells and development of full-blown diabetes.

Pathogenic Mechanisms

Recent insights into the pathogenicity of diabetes unravel inflammation as a crucial step for development of impaired insulin secretion and insulin resistance. The main mechanisms are glucotoxicity, lipotoxicity, oxidative stress, endoplasmic reticulum stress, and amyloid deposition.

Glucotoxicity: High blood sugar can impair insulin secretion and damage β cells in pancreas. Even minor fluctuations in blood glucose years before the onset of diabetes are found to have toxic effects in the β cells. This indicates the importance of maintaining a healthy diet and avoiding high glycaemic foods which produce rapid surge of blood sugar level.

Lipotoxicity: Long chain fatty acids are found to be raised in insulin resistance. This results in impaired β cell secretion and leads on to β cell death. The fat deposition in pancreas might be one mechanism of producing β cell damage.

It is also understood that free fatty acids (FFA) in combination with high glucose can potentiate diabetes. It is called as glucolipotoxicity.

Oxidative Stress: High glucose is a stressor, and there are several other factors producing stress in tissues which generate reactive oxygen species (ROS). If the pancreatic β cells are not having sufficient antioxidant enzymes to protect against these damages, it leads on to β cell death.

Endoplasmic Reticulum Stress: The endoplasmic reticulum is an important organelle in the cells which serves protein synthesis and transport. In conditions of stress like insulin resistance, there is an already high level of insulin in the system. However, since they are not utilised properly, the body tries to increase the secretion of insulin, which in turn stresses the ER.

Amyloid Deposition: Amyloids are insoluble protein fibres which are clumped in tissues. Amyloids are seen deposited in β cells of pancreas in diabetic patients. It is not clear whether the amyloid deposition results in β cell death.

Insulin Resistance:

In most of the cases, type 2 DM starts with inflammation and subsequent insulin resistance. The constant exposure to pro-inflammatory foods leads to changes in insulin sensitivity in 'target organs'. This results in insulin resistance and full-blown diabetes. Insulin resistance can be observed right from prediabetic to advanced stages of diabetes (1).

There are evidences to link inflammatory processes with metabolic disorders like obesity and diabetes. These are characterised by increased production of pro-inflammatory cytokines (IL 1β and TNF α) and infiltration of immune cells (2). Both IL-1β and TNF α are implicated in insulin resistance. There are multi-protein substances identified as 'inflammasomes', which activates IL-1β from its inactive form (3). The IL1β is produced from pancreatic beta cells and fat cells.

Free fatty acids (FFAs) alone can stimulate the release of IL-1β in islets, and its production is accelerated in the presence of high glucose levels (1). It is also found that inflammation of islet cells can impair the function of β cells in diabetes (4).

The Classical or Oft-repeated Ones

Diabetes is called as multifactorial disease as plenty of factors are shown to be responsible for its causation. The strongest among these are obesity and physical inactivity. Even though genes have a significant impact on diabetes causation, until now, we could not get breakthrough in identifying a single gene causing diabetes (5). The search is still on, and I am pretty sure that it would be difficult to nail a single specific gene for diabetic causation applicable to all different populations. It has been identified that certain chromosomal loci (such as 2q37, 15q21, 10q26 and 3p in Mexican Americans; 12q24, 1q21-1q23 in Whites; 1q21-1q23 and 11qq23-q25in Pima Indians; 1q21-1q24 in French) are more commonly associated with diabetes in certain populations, but their role is not consistently established (6). It is observed that even in identical twins, with the susceptible genes (the right kind of gene to get diabetes), diet and physical activity are strong enough to delay the onset of diabetes (or to prevent it).

The following are a few of the common causative factors of diabetes:

- Physical inactivity and obesity
- Metabolic Syndrome
- Insulin resistance
- Abnormal glucose production by the liver
- Balance between insulin and glucagon
- Genetic susceptibility
- Cell signalling and regulation
- Beta cell dysfunction

Some people are more prone to develop diabetes; the factors determine their proneness are called risk factors. These are identified based on population studies.

So people with the following characteristics are *more likely* to get diabetes:

- People above forty-five years
- Parents or siblings with diabetes
- Ethnicity: African, Asian, Hispanic
- History of delivering a baby weighing more than nine pounds
- Gestational diabetes (diabetes in pregnancy)
- Polycystic ovarian syndrome
- Acanthosis nigricans
- History of coronary artery disease
- **Overweight or obese**
- **Physically inactive**
- **Prediabetes**
- **High blood pressure**
- **Low HDL** (the so-called good cholesterol*) and high triglycerides

*There is no good or bad cholesterol. In the near future, there is a high probability of abandoning the categorisation of cholesterol into good and bad. In fact, all types of cholesterol are needed and produced by the body, but they have different roles.

Among the above-mentioned risk factors, the highlighted ones are called as modifiable risk factors, as a healthier lifestyle can modify them.

It is worth mentioning that around 80-95% of European whites with diabetes are overweight or obese at the time of diagnosis (7).

The Neglected Ones

There is no dearth of evidence for the importance of diet and exercises in lifestyle disorders. Diabetes, overweight or obesity, high cholesterol, heart disease, insulin resistance, and high blood pressure are considered as lifestyle disorders. But instead of driving down the importance for modifying the lifestyle, billions are spent in search of a magical pill for these disorders!

Let us consider two real life scenarios.

One: Imagine a person is detected to be having high fasting blood sugar (say 128 mg/dL) in his routine blood test and seeks help from a doctor. The doctor advised him to repeat the blood test at a later date. The repeat test found that his blood sugar value is in the same range. He was given a thorough physical examination to find out any complications (eye examination, nerve tests, urine for micro albuminuria). He was given advice about the importance of maintaining a normal blood sugar and how to achieve it with non-pharmacological (without pills) means. He was asked to avoid certain foods, start exercise, to keep a watch on his blood sugar and to undergo periodic check-ups with his GP.

Second: Person who got the same blood sugar values approached another doctor who believes in 'aggressive pharmacological' intervention, which is very popular nowadays. He was somewhat threatened by the doctor, saying that unless he starts some medications, he would face terrible complications, such as loss of eyesight, legs, kidney damage, sexual dysfunction, heart attack, so on and so forth. Prescriptions were given for more tests and for blood sugar, blood pressure, cholesterol, blood thinning medications. He was also referred to specialists like nephrologist, neurologist, cardiologist, and podiatrist.

The chances of the second person becoming a devout visitor of hospital are more compared with the first one as he is more stressed, alarmed, and worried about dying at any moment. Sadly, most of the times people are comfortable with these approaches as it is convenient for them to pop a pill and 'forget' about their illness. They are also happy that the doctor is giving him 'proper' attention and care. Hospital industry promotes this approach as it keeps their business running. But just popping a pill is not sufficient in managing diabetes, which mainly is a lifestyle-induced disorder.

I would like to draw your attention to the famous Diabetes Prevention Program (not many doctors are aware or mention it that often so also the pharmaceutical companies!). It was a major multi-centric trial (a study conducted in multiple centres), aimed to find if a modest weight loss through diet and increased physical activity or treating with an oral diabetic drug (named metformin) could prevent or delay diabetes in the participants. In fact, the study was comparing the efficacy of a proven powerful medicine with humble diet and exercise! They selected 3234 participants from twenty-seven centres around United States and randomly allocated to four groups. All participants were overweight and classified as pre-diabetic; means their fasting blood sugar values were higher than normal but less than adequate to label them as diabetic. The four groups were the following:

- Lifestyle Intervention Group: The participants in this group received intensive training in diet, physical activity, and behavioural modifications. By eating less and doing 150 minutes of exercise per week, this group was expected to lose 7% of their body weight.
- Metformin Group: This group was given metformin 850 mg twice a day. They were also given advice for diet and exercise but not given any motivational counselling.
- Placebo Group: This group was given a look alike inert tablet (aka placebo) but not the real one. They were also given advice for diet and exercises but no motivational counselling.
- Troglitazone Group: This group was given a new drug. But the researchers were forced to abandon their study as this medicine caused serious liver complications.

So the three groups instead of four completed the study and results were like this:

- The participants in the Lifestyle Intervention Group reduced their risk of developing diabetes by *58%*.

The result was uniform across all participants in all ethnicity, age group, and in both sex. About 5% of lifestyle group developed diabetes each year.

- Participants in metformin group reduced their risk by 31%. Do not forget that this was the combined effect of diet, exercise, and medication but without behavioural modifications. They also found that metformin was least effective in people older than forty-five years. It was most effective in age group 25-44 and people with a body mass index 35 and above. About 7.5% of metformin group developed diabetes each year.
- Eleven per cent of participants of the placebo group developed diabetes every year.

It is clearly proved from the study that lifestyle intervention was almost twice better than the 'powerful' medication in reducing the risk of developing diabetes. Yet proponents of drugs preferred to selectively report that metformin was effective in preventing diabetes risk! They conveniently neglected the lifestyle intervention! Is this bias is because of powerful pharmaceutical *intervention* or because of 'dishing what the people like' attitude?

Those interested to get more information about the DPP study can go to the link given at the end of this chapter (8).

Diabetes—an Inflammatory disease

Insulin resistance is found in around 90% people with diabetes, and it precedes the first symptom of diabetes. There is more ground to the theory that diabetes is part of a spectrum of events starting with inflammation. Raised levels of systemic inflammation have been shown to predispose to insulin resistance which is seen in early stages of type 2 DM. Studies show the raised levels of CRP and interleukin 6 (markers of systemic inflammation) are predictors for diabetes in women (9).

There is so much of confusion regarding what exactly causes diabetes. We have seen the multitudes of factors causing diabetes. I would rather try to see this from another angle. Diabetes, heart disease (coronary artery disease—CAD), obesity, arthritis, insulin resistance and related conditions, and degenerative diseases are related closely. So can't all these have a common root? Evidences and recent researches tell us there is a high probability of a common pathway. On a closer look, we see inflammation as a common factor in all these disorders. There is growing evidence for subclinical chronic inflammation as an important pathogenic factor in diabetes development (10). In fact, the theory of linking inflammation to diabetes is not new. This was known at least 100 years back when in 1876 Ebstein found out sodium salicylates would make diabetic symptoms to disappear (11).

Low-grade systemic inflammation is found in type 2 diabetes, insulin resistance, obesity, and atherosclerosis. One of the inflammatory cytokines—TNF α—plays a direct role in Metabolic syndrome. There are high levels of TNF α expressed in the skeletal muscle and plasma of patients with type 2 diabetes. In these conditions, fat (which produces TNF α) serves as the main source of circulating TNF. TNF α impairs insulin-mediated glucose uptake and has a direct inhibiting effect on insulin signalling. It also produces insulin resistance indirectly by release of free fatty acids from adipose tissue. TNF α stimulates the release of another cytokine—also from fat—IL-6 which enhances fat breakdown. IL-6 exerts its anti-inflammatory properties by regulating the levels of TNF α and stimulates the production of IL-1ra and IL-10, which are anti-inflammatory cytokines.

Now we examine what is inflammation and how does it lead on to insulin resistance.

Inflammation can be described as a response by the body's immunological mechanism to remove injurious stimuli and initiate healing process. An example is bee sting (which I think is not nice to have and not a simple mechanism!). Imagine you are stung by a bee (ouch!). It produces severe pain followed by burning, redness, itching. The sting produces attraction of immunological response followed by a chain of events (recruiting white blood cells like neutrophils, macrophages) which is manifested as inflammation. Through this inflammatory response, the body tries to contain the damage locally to remove the offending stuff from the bee sting and helps in healing process.

In most of the cases, inflammatory response serves as a protective mechanism. But sometimes, they become aggressive and destroy body's normal tissues. This altered immunological response is the reason for disorders of immune dysfunction. This can be easily explained by a simple analogy. The military and police are supposed to protect the life of civilians and to maintain law and order of a country. Imagine a situation where these forces become reckless and kill the civilians. The same situation happens in immune dysfunctions.

Most of our day-to-day activities result in triggering a long lasting low-grade inflammation, called as chronic low-grade inflammation. This chronic low-grade inflammation is considered as the reason for many of the chronic diseases in the modern world.

There are many causes for this chronic low-grade inflammation. It could be due to subclinical viral infections, foods we eat, methods of cooking our food, mental stress, atmospheric pollutants, so on and so forth.

I would like to start with food, as most of us are so naive to realise it as cause for long-standing changes in our body (apart from contributing to our ever-expanding waistline). So people resort to plenty of tasty experiments to satisfy (satiate) their hunger. Yes, very true. Many a times we sacrifice the whole of our life for the sake of an organ which is hardly four inches in length: the tongue! (12)

Here I would like to introduce one new term 'advanced glycation end products'—commonly known by its abbreviated form AGEs. While naming it nobody would have realised that AGEs would be strong enough to produce aging and degenerative changes in human body!

Glycation is a process which refers to binding of glucose or other reducing sugars to proteins. The initial reaction is between carbonyl group of glucose and amino groups of proteins. Don't bother about these terms if you are not having biochemistry background; they simply indicate the various names given to 'active end' in a 'chain' depicting its 'chemical formula'. The initial reaction between the carbonyl group and amino groups results in formation of a Schiff base, which is highly unstable. This Schiff base soon rearranges itself to a more stable substance called as Amadori product. The Amadori product undergoes a set of irreversible reaction with oxidation to a complex compound called as advanced glycation end product or AGE.

Advanced glycation end products (AGE), the simplistic definition would be 'the chemicals formed when protein or fat binds with sugar'. The AGEs either can be formed inside the body (endogenous) or outside (exogenous). When it is formed inside, it is the result of non-enzymatic process between extracellular protein and glucose. After exploring more into endogenous AGEs, we come back to our topic.

Glycation process in vivo (in tissues) can generate two kinds of products: early and advanced glycation end products. HbA1c or glycated haemoglobin is an early product (13). Diabetics would be familiar with the blood test called HbA1c, short form for glycated haemoglobin—which is a marker of diabetic control for the last 8-12 weeks. The red blood cells containing the red pigment haemoglobin has a life span of 100-120 days. The glucose molecule reacts with Hb resulting in glycated Hb, which is a non-reversible process. In people with poorly controlled diabetes, since more sugar is available for the reaction, the values of HbA1c are higher than people who are having controlled diabetes. If the HbA1c is high, it means your diabetic control over the last 1½ to 2 months was not satisfactory. The HbA1c is the result of glycation of haemoglobin. (One of the early reaction products called as Amadori product). So we can infer that the more oxidation of haemoglobin, the higher the HbA1c value. And more glucose means more HbA1c. It is also found that HbA1c is a precursor of the advanced glycation end product on Hb (AGE-Hb), and they are correlated (14).

Researchers conclusively proved HbA1c as an independent risk factor for increased serum concentration of AGEs. Higher values of serum AGEs are associated with increased severity of eye and kidney damage in diabetic people (15). It is found that AGEs are formed at a constant slow rate in normal body. It starts as early as in embryo stage and accumulates over time. The process is speeded up in diabetes (15). There are a lot of evidences suggesting AGEs as an important factor in mediation of most of the diabetic complications (16,17).

Now we see whether the AGEs formed outside the body (exogenous AGEs) can be introduced into the body. Tobacco smoke is an example of an exogenous AGE. The so-called pre-AGEs in tobacco are converted into active or toxic AGEs during burning of a cigarette. High levels of AGEs are found in the serum of smokers. An interesting and equally alarming fact is that food also serves as an exogenous AGE source! AGEs are produced *unintentionally* when we heat food at certain temperatures and combinations. Since the generation of AGEs correlate with good flavouring, some manufacturers add them in food to make it 'yummy' (*intentional!*). About 10% of ingested AGEs are absorbed with the food, and there is a direct relationship between the AGEs in circulation and AGEs consumed! (17)

When foods containing carbohydrates (or containing glucose) and proteins are heated at higher temperatures for certain duration, AGEs are formed. The main chemicals generated during AGE formation are Schiff base and Amadori products. When food is heated, it undergoes an oxidation process. This generates free radicals which are found to be potentially toxic to biological systems. Similar toxic metabolites are produced when fat (lipids) are heated, and they are called as advanced lipoxidation end products (ALEP). We know oxidation and resulting free radicals are bad to the body, so foods and supplements containing antioxidants are recommended. The ORAC value (oxygen radical absorbance capacity) helps us to understand foods which are having high antioxidant potential (13).

Around three to five times a day, our body is exposed to these chemicals in the food. But instead of trying to reduce their exposure, we resort to 'live for the moment' and rely on medications to correct the damage caused by these chemicals. Most of us are happy if our blood tests stay within the 'normal' range. Even with perfectly normal routine test results, one could still be having high levels of toxins in his system. I am pretty sure that at least some of you would be getting these questions in your mind. 'Which is the best lab test for detecting these toxins?' 'Where is it available?'

This attitude is just like the mind-set of a kid sitting for an exam without preparation. He is more worried about the grades and its consequences than lack of preparation for the exams.

By adopting a diet low in AGEs, we can bring down the circulating AGE levels at least by 30% per month without showing significant changes in HbA1c (16).

Low-Grade Inflammation and Innate Immunity

Innate immunity is our body's first line of defence against invading organisms. These are mediated by different cells known as macrophages (a type of white blood cells), antigen presenting B cells, dendritic cells, endothelial cells, intestinal epithelial cells, Kupffer cells in the liver and adipocytes.

Inflammation is a local response which is protective (most of the times) against injury. But there is a systemic response in the body which is part of inflammation known as the acute phase response. This response is manifested by certain circulating proteins called as acute phase reactants. C-reactive protein (CRP), fibrinogen, and complement are some of the examples for acute phase reactants. These are produced in the liver and help to contain the injury or help in healing.

The innate immunity and acute phase response are linked with the nervous system mainly through the HPA (hypothalamo-pituitary-adrenal axis). So acute phase response can also result from psychological stress.

Inflammation Caused by Virus

The recent reports of viral infection as a cause of type 2 diabetes are interesting. One particular type of virus called as cytomegalovirus (CMV) infection is considered as a pathogenic factor in producing type 2 diabetes. Some studies have mentioned that there is a high prevalence of high fasting glucose and HbA1c levels in CMV seropositive older adults (18).

Hepatitis virus C infection is reported to be higher in type 2 diabetes. There is up to tenfold increase in diabetes reported in HCV seropositive patients compared with chronic liver disease patients. But it is not clear whether the hepatitis infection makes the person susceptible for diabetes (19).

Fructose and Metabolic Syndrome

Fructose is a type of sugar (monosaccharide) present in fruits and in normal sugar (sucrose). It is a common ingredient in many foods we consume on a daily basis. When it is extracted from corn, it is known as high-fructose corn syrup (HFCS) or simply corn syrup. Clinical studies have shown fructose intake to induce high triglycerides, high uric acid, weight gain, diabetes, and increased blood pressure (20).

Fructose Metabolism: Fructose is absorbed unchanged in human small intestine. It is then taken to the liver where more than 50% of it is metabolised. The rest of it is metabolised in kidneys and fat tissues. Under normal conditions, fructose is converted to fructose 1 phosphate by an enzyme called as fructokinase. The conversion of fructose by the enzyme results in low ATP inside the cell. The low ATP causes uric acid accumulation, which is a toxic product.

High intake of refined sugars increases the incidence of insulin resistance. Junk foods made up of refined carbohydrates, such as refined flour, sucrose, and high fructose syrup are considered as a major contributor for obesity, systemic inflammation, and metabolic syndrome. They are called as empty calories as they are devoid of minerals and vitamins. Studies in fructose overfed rats show a rapid increase in reactive oxygen species (ROS), which are toxic to tissues. The ratio of vitamin E to triglycerides is also found to be low in these rats. This favours lipid peroxidation of lipoproteins, which in turn can lead on to heart complications. The inflammatory response and oxidative stress are important in development of insulin resistance in high fructose-fed rats (21).

Fat-Induced Inflammation

Insulin resistance is the key metabolic abnormality in type 2 diabetes, and obesity is the most common cause of insulin resistance (24). Diets rich in saturated fats are linked with increased obesity and type 2 diabetes. Saturated fats in the diet are absorbed from intestine and get into chylomicrons. From there, it enters into lymphatic circulation and finally reaches the blood circulation. These dietary SFA act as powerful inflammatory mediators to trigger inflammatory pathways in innate immune cells like macrophages and insulin target cells like fat, muscle, and liver cells.

Human adipose tissue has got many roles. Contrary to the widely accepted concept of it as a storage organ, the new evidences indicate a more dynamic role. The white adipose tissue (WAT) acts as an endocrine organ which has secretory functions. WAT secrete factors needed for balancing glucose, energy metabolism, food intake and body weight regulation, haemostasis (stopping bleeds) and immune function (22). When the size of fat tissue increases, the size and number of the adipocytes (fat cells) also increase. The expanded fat cells secrete inflammatory substances, which in turn result in systemic inflammation.

The fat-filled adipocytes finally die releasing their contents, which are inflammatory in nature.

Adipose tissue or plainly known as fat tissue stores excess energy as fat. In a way, it prevents lipid accumulation in other body organs. But when more fat is accumulated in the adipose tissue, it results in stress signals and deranges its metabolic function.

Sirtuin 1 (Sir T1)—a nutrient sensing histone deacetylase—is found to be reduced by overfeeding. The SirT1 helps in preventing inflammatory mediators release in fat tissue (by blocking entry of macrophage and subsequent release of inflammatory mediators). It has been shown that a high-fat diet triggers inflammation by reducing the level of SirT1 which normally protects against it (23,24).

It is also worth mentioning that apart from the high energy content of fats, the quality of dietary fat is important in producing insulin sensitivity (25).

High levels of IL-6 and TNF-α in patients with the metabolic syndrome are associated with truncal fat mass. Both TNF and IL-6 are found to be produced in adipose tissue.

- Mechanism of diabetes is not simple.
- Nobody becomes diabetic overnight.
- Diabetes is an inflammatory disease.
- Insulin resistance is a major step in metabolic pathway.
- Diet and lifestyle have a significant effect on inflammation.
- Consumption of high GI foods is not the only reason for diabetes.
- Fats are important to our body; but modification of fats results in bad effects.
- Cooking methods can result in toxin formation (AGEs) and can result in diabetes.

References

1. Böni-Schnetzler M, Donath MY. Increased IL-1β activation, the culprit not only for defective insulin secretion but also for insulin resistance? Cell Res. 2011;21:995-7; PMCID: PMC3193499

2. Donath MY, Shoelson SE. Type 2 diabetes as an inflammatory disease. Nat Rev Immunol. 2011;11(2):98-107; PMID: 21233852

3. Martinon F, Burns K, Tschopp J. The inflammasome: a molecular platform triggering activation of inflammatory caspases and processing of proIL-beta. Mol Cell. 2002;10(2):417-26; PMID: 12191486

4. Donath MY, Böni-Schnetzler M. Islet inflammation impairs the pancreatic β-cell in type 2 diabetes. Physiology. 2009;24:325-31; PMID: 19996363

5. van Tilburg J, van Haeften TW, Pearson P, Wijmenga C. Defining the genetic contribution of type 2diabetes mellitus. J Med Genet. 2001;38:569-78

6. Vionne N, et al. Genomewide search for type 2 diabetes—susceptibility genes in French Whites: evidence for a novel susceptibility locus for early-onset diabetes on chromosome 3q27-qter and independent replication of a type 2—diabetes locus on chromosome 1q21—q24. Am J Hum Genet. 2000;67:1470-80

7. Astrup, Healthy lifestyles in Europe: prevention of obesity and type II diabetes by diet and physical activity. Public Health Nutr. 2001;4(2):499-515

8. http://www.diabetes.niddk.nih.gov/dm/pubs/preventionprogram/index.aspx#results

9. Pradhan AD, Manson JE, Rifai N, Buring JE, Ridker PM. C-reactive protein, interleukin-6, and risk of developing type 2 diabetes mellitus. JAMA. 2001;286:327-34

10. Ake Sjoholm, Thomas Nystrom. Inflammation and the etiology of type 2 diabetes. Diabetes Metab Res Rev. 2006;22:4-10

11. Shoelson SE, Lee J, Goldfine AB. Inflammation and insulin resistance. J Clin Invest. 2006;116(7):1793-801

12. http://en.wikipedia.org/wiki/Tongue

13. http://en.wikipedia.org/wiki/Oxygen_radical_absorbance_capacity

14. Turk Z, Mesić R, Benko B. Comparison of advanced glycation end products of haemoglobin (Hb-AGE) and haemoglobin A1c for the assessment of diabetic control. Clin Chim Acta. 1998;277(2):159-70

15. Aso Y, Inukai T, Tayama K, Takemura Y. Serum concentrations of advanced glycation end products are associated with the development of atherosclerosis as well as diabetic microangiopathy in patients with type 2 diabetes. Acta Diabetol. 2000;37(2):87-92

16. Doreen Hsu , Victoria Zimmer. Glucose, advanced glycation end products, and diabetes complications: what is new and what works. Clin Diabetes. 2003;21(4):186-7

17. Jaleel A, Halvatsiotis P, Williamson B, Juhasz P, Martin S, Nair KS. Identification of amadori-modified plasma proteins in type 2 diabetes and the effect of short-term intensive insulin treatment. Diabetes Care. 2005;28(3):645-52

18. Sijia Chen , Anton JM de Craen, Yotam Raz, Evelyna Derhovanessian, Ann Vossen CTM, Rudi Westendorp GJ, Graham Pawelec Andrea B Maier . Cytomegalovirus seropositivity is associated with glucose regulation in the oldest old. Results from the Leiden 85-plus Study. Immun Ageing. 2012;9:18

19. Mason A. Viral induction of type 2 diabetes and autoimmune liver disease. J Nutr. 2001;131(10):2805S—08S

20. Schulze MB, Manson JE, Ludwig DS, Colditz GA, Stampfer MJ, Willett WC, et al. Sugar-sweetened beverages, weight gain, and incidence of type 2

diabetes in young and middle-aged women. JAMA. 2004;292:927-34

21. Rayssiguier Y, Gueux E, Nowacki W, Rock E, Mazur A. High fructose consumption combined with low dietary magnesium intake may increase the incidence of the metabolic syndrome by inducing inflammation. Magnesium Res 2006;19(4):237-43

22. Kennedy A, Martinez K, Chuang CC, LaPoint K, McIntosh M. Saturated fatty acid-mediated inflammation and insulin resistance in adipose tissue: mechanisms of action and implications. J Nutr. 2009;139:1-4

23. Chalkiadaki A, Guarente L. High fat diet triggers inflammation-induces cleavage of SIRT1 in adipose tissue to promote metabolic dysfunction. Cell Metab. 2012;16(2):180-8

24. Gillum MP, Kotas ME, Erion DM, Kursawe R, Chatterjee P, Nead KT . SirT1 regulates adipose tissue inflammation. Diabetes. 2011;60(12):3235-45

25. Osborn O Sears DD, Olefsky JM. Fat-induced inflammation unchecked. Cell Metab. 2010;12(6):553-4

CHAPTER 3

An Ayurvedic Perspective

The concept of diseases causing urine sugar (glycosuria) is not new. This entity is mentioned long back in 'Vedas', the sacred literatures of Indian civilisation (dates back to 2000-500 BC or even older as initially they were transmitted by oral lore). Ayurveda is considered as sub-Veda of *Atharvaveda*. In Atharvaveda, a condition named 'Asrava' is mentioned. It is considered as a general term encompassing urinary problems manifested as profuse, sweet, or turbid urine.

The name *Prameha* was fist mentioned in *Caraka samhita* (medical compendium by Caraka, which is supposed to be compiled in textual format around 600BC) (1) and in other major text books of Ayurveda like Susrutha Samhita (Surgical Compendium by Susrutha) (2) and Madhava Nidana (Compendium of Etiologies of Diseases by Madhava) (3). Twenty different types of Prameha are mentioned in Ayurvedic text books. One particular type among the different Pramehas mentioned, Madhumeha, is more akin to diabetes mellitus.

The following section can be skipped, if you are not interested to know about it in detail.

Classification of Prameha

There are different types of classification mentioned in text books for Prameha.

- **Hetu Bheda (As per Aetiology)**

Sahaja (Congenital)

Kulathaja (Familial)

Apathyaja (Due to improper diet and lifestyle)

- **Mutra Vikara Bheda (As per abnormalities in urine) Twenty types**

Khapjaja: Ten types

Pithaja: Six types

Vathaja: Four types, which include *Madhumeha*

All these twenty types are not same as type 2 diabetes. I think there is a significant overlap of different urinary abnormalities mentioned as Prameha in Ayurveda. All scholars agree that *Madhumeha* can be considered as Diabetes Mellitus of the modern era. In some texts, prameha and madhumeha are used interchangeably. Even though some scholars consider madhumeha as diabetes mellitus (DM), I consider inappropriate for us to compare an old scientific classification with a modern one. The so-called metabolic syndrome fits in somewhere among the spectrum of disorders mentioned under the umbrella term of 'prameha' or 'madhumeha'. My aim of this chapter is not to explore the veracity of prameha and its correlations. Since such discussion becomes voluminous, I restrict myself by pointing out that a disorder similar to diabetes mellitus was prevalent in those days. The scholars in those days also acknowledged it mainly as a disorder of inactivity and overindulgence. The following points have to be kept in mind before we delve more into this topic.

1. The Ayurveda is an old medical system which is at least 5,000 years old, and many of its diagnostic methods are based on symptomatology. They are not based on any laboratory investigations as in modern medicine.

2. There is a diagnostic criterion for type 2 diabetes in modern medicine, which is based on blood test (fasting and postprandial blood sugar). These kinds of tests were not available 5,000-10,000 years back.

3. Recent evidences force us to consider type 2 diabetes as part of metabolic syndrome than considering it as a separate entity. We are also aware of other medical conditions linked with type 2 diabetes or highly prevalent in the diabetic population. Obesity, polycystic ovarian disease, degenerative disorders, cancers, heart diseases are some examples. No longer it is wise to consider diabetes as a stand-alone entity but as a manifestation of ongoing metabolic abnormality happening in the body. This clustering gives us a plausible explanation for including different conditions under the broad category of Prameha.

As per Ayurvedic philosophy, any imbalances in the 'doshas'(principal forces) would ultimately result in disorders, and diabetes is not an exception. In Ayurvedic literature, the aetiology is classified broadly under two categories: Prenatal Factors (hereditary factors and factors during pregnancy) and Post natal factors (acquired factors).

Pre-Natal Factors

Hereditary Factors

Bija Dosha: Defects in the egg or sperm can result in genetic predisposition for diabetes. Ayurveda mentions that excessive indulgence of sweet foods and drinks can result in this defect. Ayurveda considers sugar, rice, breads, milk, and butter as a risk of diabetes causation. We know foods with high glycaemic index can cause metabolic derangements, insulin resistance, and diabetes. Protein in bovine milk (more of A-1β casein) is proven to cause autoimmune destruction of pancreatic β cells and type 1 diabetes. It is found that the β casein antibody levels are high in latent autoimmune diabetes in adults (LADA) (4). In a Finnish study among siblings of diabetic kids, those who take more than 540 ml of milk per day were found to have five times the risk to become diabetic. High-fat foods are known to produce lipotoxicity of pancreas and subsequent diabetes causation (5).

Intrauterine Factors

As per Ayurvedic concept, mother's diet and activity levels are determinant factors for diabetes in the progeny. It is widely acknowledged that gestational diabetes is associated with high birth weight babies. There are high incidences of obesity among these kids when they attain adolescence (6). The metabolic effects of intrauterine environment are same irrespective of diabetic status of mother (Gestational diabetes, type 1 or type 2) (7).

The prenatal factors are designated as *Sahaja/Jathaja* by different authors in Ayurvedic text books, and it is assumed comparable to genetic/hereditary factors as per modern classification. It is not wise to consider all these as causative for type 1 as they also can result in type 2 diabetes.

Post-Natal (Acquired Factors)

They are mentioned as 'apathyanimittaja' in Ayurvedic texts and represents exclusively type 2 diabetics who are obese. These are diet, lifestyle, and psychological factors.

Diet: These are excess intake of yogurt, meat, milk, newly harvested grains, foods which are sweet (mainly sugar and molasses), cold, oily, and difficult-to-digest foods. We have already seen the link between bovine milk protein and diabetes, especially type 1 diabetes. High fat, especially animal fat, has been indicated as a mechanism for reducing insulin sensitivity by altering the fatty acid composition of membrane lipids. The cell membrane (covering of the cell) has got plenty of fat cells in it, known as membrane lipids. These membrane lipids serve as attachment sites for various substances. The attachment triggers a chain of reactions which ultimately determine the protein synthesis and other nuclear activities in the cell. There is active debate on whether the saturated fats or the polyunsaturated ones are causing the damage. Fats per se are not bad to the body. The body needs certain quantities of good fats for its day-to-day activities, such as hormone production, absorption of fat soluble vitamins. But it is understood that the so-called *trans fats* are associated with type 2 diabetes development (8).

The link between high GI foods and diabetes is already mentioned.

Lifestyle Factors

Ayurveda mentions sedentary lifestyle, prolonged sitting, excessive sleeping, day-time sleepiness, lack of exercise among the causes of diabetes. Lifestyle is one of the most important factors that influence diabetes in the modern world. It is now acknowledged even in kids as a major contributing factor for diabetes. It is sad to see that more kids spend their leisure time in computers and televisions rather than playing outdoors. This lifestyle, in combination with snacking of high GI, trans-fat laden foods is a perfect recipe for diabetic causation.

Psychological Factors

Stress is another important factor in diabetes in susceptible people. It works on the hypothalamo pituitary adrenal axis (HPA axis) and produces a lot of hormonal changes in the body. It has been shown that even a single incidence of severe stress is sufficient enough to produce diabetes in susceptible people. Ayurveda gives a lot of importance to the role of mind in various disease pathogenesis. It explains the importance of mind-body interaction with a simple analogy of butter in a heated vessel: the butter is affected by the heat conducted from the vessel (even if the butter is cold). Similarly a restless mind can affect the body; so also a restless body affects the mind.

Stress due to depression, environmental factors, or perceived stress results in increased release of corticosteroids and other neurohormonal factors. This in turn can lead on to abdominal obesity, insulin resistance, and other features of metabolic syndrome (9).

In depressed individuals, there is an increased activity in hypothalamo-pituitary-adrenal (HPA axis) which results in increased cortisol secretion and the resulting glucose intolerance. This is partly due to the insulin opposing action of cortisol and resulting insulin resistance. Increased cortisol levels can also cause central obesity and hypertension (10,11).

Categorisation as per body habitus

Caraka Samhita classifies Prameha mainly into two as per body habitus.

Krisha Pramehi: Lean diabetic (Type 1?)

Sthoola Pramehi: Obese diabetic (Type 2?)

Categorisation as per prognosis

Sadhya: Easily amenable to treatment/Curable

Yapya: Controlled with management

Asadhya: Difficult to control/Incurable

Prodromal Symptoms: *Poorvarupa*

In ancient times too much of importance was given to clinical symptomatology. This was essential as there were no laboratory tests to determine the disease. The only way to pick up diseases in their early stage was with prodromal features. The following are the prodromal symptoms mentioned for Prameha. Excessive sweating, body odour, laziness, preference for rest, smearing (? coating) in the eyes, tongue, and ears, heaviness of body, excessive growth of hair and nails, thirst, sweetness of mouth, numbness, and burning sensation of hands and feet, ants and bees attracted to urine and body.

As you can see the majority of the above-mentioned symptoms are associated with diabetes either in its early or late stage.

- Disorders like diabetes/metabolic syndromes were prevalent in olden times.
- Ayurveda recognized importance of genetic, environmental, psychological, and dietary factors in their causation.
- There were in-depth descriptions of symptomatology of these disorders in olden days.
- Most of these correlate with the known symptoms and complication profile of 'diabesity'.

References

1. Caraka samhita; Chikitsa Sthana; Chapter 6
2. Susrutha Samhita; Nidana Sthana; Chapter 6/Chikitsa Sthana; Chapters 11, 12 and 13
3. Madhava Nidana; Chapter 33
4. Monetini L, et al. Antibodies to bovine beta-casein in diabetes and other autoimmune diseases. Horm Metab Res. 2002;34(8):455-9
5. Robertson RP, Harmon J, Tran PO, Poitout V. Beta-cell glucose toxicity, lipotoxicity, and chronic oxidative stress in type 2 diabetes. Diabetes. 2004;53: 119-24(Suppl 1)
6. Gillman MW, et al. Maternal gestational diabetes, birth weight, and adolescent obesity. Pediatrics. 2003;111(3):e221-6
7. Dabelea D. The presdisposition to obesity and diabetes in offspring of diabetic mothers. Diabetes Care. 2007;30(Suppl 2):S169-74

8. Hu FB, Van Dam RM, Liu S. Diet and risk of type 2 diabetes: the role of types of fat and carbohydrate. Diabetologia. 2001;44(7):805-17

9. Vaccarino V, Bremner JD. Stress response and the metabolic syndrome: Hospital Physician Board Review Manual. Cardiology. Vol 11(Part 2)

10. Andrews RC, Walker BR. Glucocorticoids and insulin resistance: old hormones, new targets. Clin Sci. 1999;96:513-23

11. Rosmond R. Stress induced disturbances of HPA axis: a pathway to type 2 diabetes? Med Sci Monit. 2003;9(2):RA35-9

CHAPTER 4

Complications of Diabetes

- Hypoglycaemia (Low sugar)
- Hyperglycaemia (High sugar)
- Heart and blood vessel disease
- Stroke
- Nerve damage
- Eye complications (Retinopathy)
- Kidney damage (Nephropathy)
- Foot problems
- Sexual dysfunction
- Osteoporosis
- Alzheimer's disease
- Skin problems
- Infections
- Muscle, joint issues

We need to address the following questions when considering complications of diabetes.

Why there are so many complications in diabetes?

What are the mechanisms in which they are produced in the body?

Are there any particular reasons for its predilection to certain body tissues?

In the following sections, we try to find out answers for these questions.

Most of the cells in the body need insulin for glucose entry. But the cells of retina, kidney, and nerves are exceptions. As they do not need insulin as an ID to get into them, the glucose has a direct entry! Considering their vulnerability of not able to sustain long without sugar, nature might have given them this privilege. But in diabetes, these privileges can be detrimental as they don't work for their favour.

High blood glucose should be having an equal effect in all body tissues but, in reality, that is not the case. Some tissues are more sensitive to hyperglycaemic damage—such as cell linings in the blood vessels of retina (our sensor in the eye) called as capillary endothelial cells, cells in kidneys known as mesangial cells in the glomerulus, nerves in the periphery known as neurons and Schwann cells. Even in situations of high blood sugar, most of the cells are able to regulate their sugar levels. But somehow the above-mentioned cells are not able cope with it and results in metabolic damage.

I warn you that the following section is going to be a bit technical. Unless you are not bothered about the technical jargon, or else you having an inquisitive mind, better to avoid it altogether.

The most commonly mentioned mechanisms are the following (1,2)

- Polyol Pathway
- AGEs
- PKC (protein kinase C activation)
- Products of hexosamine pathway

Polyol Pathway (Sorbitol Aldose Reductase Pathway)

Under normal glucose conditions, only a small fraction of glucose is metabolised through the polyol pathway. In these conditions, the enzyme *aldolase reductase* converts the toxic aldehydes to inactive alcohols (1). But in hyperglycaemia, metabolism of glucose through polyol pathway is increased. The enzyme converts the glucose to sorbitol, and the sorbitol is converted into fructose. For the step of converting glucose to sorbitol, the enzyme needs NADPH as a cofactor, which otherwise would have utilised for producing an antioxidant named glutathion. Since the NADPH is utilised for the above step, level of antioxidant is significantly reduced and the cells undergo oxidative damage known as oxidative stress.

When the blood sugar levels are normal, the enzyme *aldose reductase* has a low interest (affinity in biochemical jargon) to glucose, but increases when glucose concentration is high. So in high blood sugar situations conversion of glucose to sorbitol is accelerated which utilises a lot of NADPH. This in turn reduces glutathion levels; which is an antioxidant. In addition, it also results in low levels of nitric oxide, myo-inositol, and taurine, which are important in the detoxification mechanisms of the body.

AGEs in the Cell

AGEs in the cell can cause damage by multiple means. One of the mechanisms is by modifications of protein which are responsible for gene transcription (gene coding). When AGEs are outside the cell, they modify the extracellular matrix and impair signalling between it and cell. This results in cell dysfunction. AGEs on entering the circulation can make changes in albumin (a blood protein). When the altered albumin comes in contact with AGE receptors, they activate them to produce inflammatory substances (1).

PKC

When the blood glucose level in the cell is high, it activates the production of protein kinase C (PKC). PKC leads on to a chain of events resulting in a lot of metabolites which are harmful to the cell. (*decreased endothelin nitric oxide synthase, increased vasoconstrictor endothelin 1, transforming growth factor beta, plasminogen activator inhibitor, to name few*) (1,2). All these chemicals produce narrowing of blood vessels, blood-flow abnormalities, reduces the permeability of smaller blood vessels (capillary permeability), which ultimately result in cell death.

Hexosamine Pathway Products

When there is excess sugar inside the cell, most of it is metabolised through glycolysis. But some of the intermediaries like glucose 6-phosphate is diverted to glucosamine 6-phosphate (by *an enzyme Glutamine:Fructose-6-Phosphate Amidotransferace (GFAT)*. The end product is UDP, which can make changes in gene expression. The excess sugar shunted through this pathway would result in increased transforming growth factor beta 1 (TGFβ1), plasminogen activator inhibitor type 1 (PAI-1) which produce tissue injury.

Activation of the hexosamine pathway leads to reduction of pancreatic β cells by means of oxidative stress (3) and insulin resistance.

Apart from the above-mentioned complications, there are other issues caused by diabetes in bones and muscles. They are categorised as Musculoskeletal complications.

Now let us have a closer look into the musculoskeletal (muscle and bone) complications in diabetes (4,5).

• Hand Complications:

Stiff-hand syndrome—This is a condition resulting in the inability to straighten the fingers. This is seen in up to 50% of type 1 and also in type 2 diabetics.

Trigger fingers—Snapping of fingers when they are straightened from the bent position. This is due to swelling of tendon, which then produces locking as they pass through a tunnel in the fingers.

Dupuytren's contractures—Shortening of soft tissues in the palmar side of fingers, which results in the inability to straighten the fingers. This is seen in around 42% of diabetics.

Carpal tunnel syndrome—A condition producing increased tension in wrist region which results in nerve dysfunction. This is seen in up to 20% of diabetics.

- Shoulder Issues:

Frozen shoulder—Inability to move shoulder for over-head activities, most of the times associated with pain and stiffness. Seen in up to 19% of diabetic cases.

Calcific periarthritis—Calcium deposit around the shoulder capsule.

Reflex sympathetic dystrophy—A condition producing pain and dysfunction in hands and fingers.

- Feet Complications:

Charcot joint—Joint damage. It is a rare complication seen in only 0.4% cases.

- Muscles:

Diabetic muscle infarction—Damage to muscles due to lack of blood supply.

Proximal muscle weakness—Weakness of muscles of the upper part of arm and leg.

- Conditions Affecting the Skeleton:

Diffuse idiopathic skeletal hyperostosis (DISH)—A condition producing pain and stiffness of back. It is because of bony hardening of ligaments where they attach to spine.

Osteomyelitis/Septic arthritis—Infections to bone or joints.

Osteoporosis—A condition producing brittleness of bones.

Osteoarthros, is (OA)—Wear and tear resulting in pain and swelling of joints.

Osteoarthrosis (OA)

It is a very common condition in the general population but seen more in diabetics. Findings suggest that there is a significant increase in metabolic syndrome in persons with OA than patients without it. Similarly fasting glucose levels are higher in OA patients than in non-OA. In a longitudinal study on 927 people, which started in 1990 and continued till 2010, it was found that diabetes is a strong predictor for development of severe OA, and long-standing diabetes is detrimental for knee and hip joints (6).

Insulin-like growth factor 1 (IGF-1) helps the build-up of cartilage cells called as chondrocytes. When blood sugar is high, it leads on to IGF-1 resistance, which in turn results in less cartilage formation at damaged sites (7). Diabetic vessel disease (micro and macro angiopathies) contribute to the development of OA by influencing synovial tissue and bone beneath the cartilage. Non-enzymatic glycation of collagen—which influence the functional properties of cartilage—might also be an important factor in its development (8).

- Diabetes affects all systems of the body.
- Inflammation, high blood glucose levels, and AGEs are main causes for complications.
- AGEs can modify gene coding and affect communication between cell and its surroundings.
- High blood glucose reduces the anti-oxidants which protect the cells against damage.
- Some tissues have more predilection for diabetic complications as they have direct access to glucose.

References

1. Brownlee M. The pathobiology of diabetic complications: a unifying mechanism. Diabetes. 2005;54(6):1615-25

2. Brownlee M. Biochemistry and molecular cell biology of diabetic complications. Nature. 2001;414(6865):813-20

3. Kaneto H, Gang Xu, Ki-Ho Song, Kiyoshi Suzuma, Susan Bonner-Weir, Arun Sharma, Gordon C. Weir. Activation of the hexosamine pathway leads to deterioration of pancreatic β-cell function through the induction of oxidative stress. J Biol Chem. 2001;276:31099-104

4. Kim RP, Edelman SV, Kim DD. Musculoskeletal complications of diabetes mellitus. Clin Diabetes. 2001;19(3):132-5

5. Lawrence HW, Randy JF. The musculoskeletal effects of diabetes mellitus. J Can Chiropr Assoc. 2006;50(1):43-50; PMCID: PMC 1839979

6. Georg Schett, Arndt Kleyer, Carlo Perricone, Enijad Sahinbegovic, Annamaria Iagnocco, Jochen Zwerina et al. Diabetes is an independent predictor for severe osteoarthritis. Diabetes Care. 2012; 36(2): 403-9 PMID: 23002084

7. Kelley KM, Johnson TR, Ilan J, Moskowitz RW. Glucose regulation of the IGF response system in chondrocytes: induction of an IGF-I-resistant state. Am J Physiol. 1999;276:R1164-71; PMID: 10198399

8. Reiser KM. Nonenzymatic glycation of collagen in aging and diabetes. Proc Soc Exp Biol Med. 1991;196:11-29; PMID: 1984239

CHAPTER 5

Current Management Principles

Treatment for type 2 diabetes is mostly symptomatic. It is limited to monitoring fasting and postprandial blood sugars and adjusting the dosage of medications accordingly. Rarely patients are educated about the importance of strictly adhering to the diet and exercises. This kind of reductionist approach would lead on to addition of number of medications, and it is not uncommon for patients to land up with at least half a dozen prescription pills in a few years. Many patients mention that their appetite is low, and they don't feel like eating after having many pills. Generally the following medications are prescribed:

- One or two anti-diabetic drugs with or without insulin injection (depending on the blood sugar control)
- One or two blood-thinning medications (like aspirin and clopidogrel)
- Anti-cholesterol medications (like statins)
- Anti-hypertensives or ACE inhibitors

Below is the list of possible additions for treating complications:

- Bowel softeners for constipation

- Gastric acid suppressants (Proton-pump inhibitors or H2 blockers)
- Vitamin B preparations or Gabapentins or Tegretol for neuropathy
- Pain killers or NSAIDs for arthritis. Most of the diabetics develop degenerative changes (Osteoarthrosis) in their knee joints. Hyperglycaemia compromises synthesis of type 2 collagen and increases productions of reactive oxygen species, both are important in the development of OA (1).
- Calcium supplementations for osteoporosis

Blood Glucose Monitoring

The long-term complications of diabetes are linked with average blood glucose levels. It has been shown that maintaining a healthy glucose levels is important to prevent or delay diabetic complications. It is always desirable to have a normal blood sugar levels by following healthy lifestyle and eating habits rather than achieving it through medications. Prescription drugs should be considered only in those who fail to control blood sugar by these approaches.

I have seen people finding solace in having foods in high glycaemic index and then try to wash out the 'sin' with an extra dose of pill or insulin jab! Some people go for less guilty alternatives like artificial sweeteners or fructose. They believe in maintaining a normal test result; whatever happens in-between is not a concern for them.

Many of the artificial sweeteners are not so naive. Most of them have demonstrated their potential toxicity (at least in animal studies). We explore this in detail in a later section.

HbA1c (Glycated haemoglobin)

Glycated or glycosylated haemoglobin (HbA1c) measures the blood glucose control over the lifespan of Red blood cells (RBC), which is usually 100-120 days. So HbA1c reflects the average plasma glucose over the last 8-12 weeks.

HbA1c does not have the day-to-day fluctuations as blood glucose, and it can be performed at any time of the day. It does not require any special preparation, such as fasting. HbA1c has now been recommended by an International Expert Committee and by the American Diabetic Association as a means to diagnose diabetes.

An HbA1c of 6.5% is recommended as the cut point for diagnosing diabetes. But a value less than 6.5% does not exclude diabetes diagnosed using glucose tests.

A report published in 2009 by an International Expert committee recommended that HbA1c can be used to diagnose diabetes. The committee mentions if HbA1c is above 6.5%, a diagnosis of diabetes can be made. But it should be confirmed by a repeat HbA1c test, unless clinical symptoms and plasma glucose levels >11.1 mmol/l (200 mg/dL) are present in which case further testing is not required. HbA1c levels just below 6.5% may indicate the presence of intermediate hyperglycaemia. The precise lower cut-off point for HbA1c has yet to be defined. The American Diabetes Association suggests HbA1c values between 5.7% and 6.4% as the high-risk range. The International Expert Committee recommended that persons with HbA1c level between 6.0 and 6.5% are at high risk and can be considered for diabetes prevention interventions.

Factors influencing haemoglobin can affect the value of HbA1c. Following are the factors that influence HbA1c and its measurement (2): (This section is also bit technical and can be skipped if required.)

1. Erythropoiesis (Red blood cell production)
   - *Increased HbA1c: Deficiency of iron, vitamin B12, reduced erythropoiesis.*
   - *Decreased HbA1c: on preparations like erythropoietin, iron, vitamin B12, reticulocytosis, chronic liver disease.*

2. Altered Haemoglobin
   - *Genetic or chemical alterations in haemoglobin: haemoglobinopathies, HbF (Faetal Hb), methaemoglobin may increase or decrease HbA1c.*

3. Glycation
   - *Increased HbA1c: alcoholism, chronic renal failure, decreased intraerythrocytic pH.*

- *Decreased HbA1c: aspirin, vitamin C and E, certain haemoglobinopathies, increased intra-erythrocyte pH.*
- *Variable HbA1c: genetic determinants.*

4. Erythrocyte destruction
   - *Increased HbA1c: increased erythrocyte lifespan: Splenectomy.*
   - *Decreased A1c: decreased erythrocyte lifespan: haemoglobinopathies, splenomegaly, rheumatoid arthritis, drugs such as antiretrovirals, ribavirin, and dapsone.*

5. Assays
   - *Increased HbA1c: hyperbilirubinaemia, carbamylated haemoglobin, alcoholism, large doses of aspirin, chronic opiate use.*
   - *Variable HbA1c: haemoglobinopathies.*
   - *Decreased HbA1c: hypertriglyceridemia.*

WHO Recommendation 2011 (3)

An expert committee by the World Health Organization (WHO) recommended that HbA1c can be used as a diagnostic test for diabetes. But a rider is added that there should be stringent quality assurance tests in place. Assays are to be standardised to meet the criteria set by the international reference values and conditions which preclude its accurate measurement should not be present.

WHO recommends HbA1c of 48 mmol/mol (6.5%) as the cut-off point for diagnosing diabetes. Values less than 48 mmol/mol (6.5%) do not exclude diabetes diagnosed using glucose tests.

Fructosamine Test

Fructosamine is a compound formed as a result of reaction between carboxyl group of glucose and amino group of a protein. Fructosamine is formed from serum proteins like albumin and known as glyclated serum protein (GSP). Like HbA1c, it is used to determine plasma glucose concentration over a period but is shorter than HbA1c.

Fructosamine is not affected by disorders of red blood cells; whereas it normally affects the HbA1c values. It also has an advantage of accurately reflecting short-term changes in glycaemia—which usually corresponds to half life of Albumin, a plasma protein. Now it is recommended in all patients with red blood cell disorders and in those having disparities between blood glucose and HbA1c measurements.

- Current concept of management is a reductionist approach.
- Restricted to 'constant observation' than prevention.
- Blood glucose values are transient.
- Needs better parameters for glucose control and complication prevention.
- HbA1c is a better indicator of diabetic control, and it shows the last three months' sugar status.
- Special test like fructosamine is needed for people who are having red cell disorders.

References

1. Berenbaum F. Diabetes-induced osteoarthritis: from a new paradigm to a new phenotype. Ann Rheum Dis. 2011;70(8):1354-6

2. Gallagher EJ, Le Roith D, Bloomgarden Z. Review of hemoglobin A1c in the management of diabetes. J Diabetes. 2009;1:9-17

3. Use of glycated haemoglobin (HbA1c) in the diagnosis of diabetes mellitus. Abbreviated report of a WHO consultation: 2011;WHO/NMH/CHP/CPM/11.1

CHAPTER 6

Diet and Diabetes

Volumes of research papers are available linking diabetes with poor eating habits.

Until recently, a bad diet was believed to cause changes in the blood cholesterol or sugar and refraining from fats and sugar were recommended mainly for people with these conditions. There were no other reasons for having a 'healthy diet', as its beneficial effects were not scrutinised by research methodology. Researches in this field point towards long-lasting benefits of food on the body. Its effects are far beyond than just controlling cholesterol and blood sugar. It is worth recollecting the findings of Diabetes Prevention Program which I mentioned earlier in this book.

For centuries, humans adopted a food pattern which was available locally, suitable to their race and culture. However, globalisation and acculturation has resulted in a lot of give and take between different cultures and races. As a result, cultural differences are being reduced slowly and significantly, thanks to the development of transportation, multi-cuisine restaurants and migration of people for work and better living standards.

Cavemen had totally different food habits, known as the Paleo diet or hunter-gatherer diet. Humans are believed to have followed this diet for nearly 2.5 million years and this determined the prevalent food habits as far back as 10,000 years. Their diet consisted mainly of fish, grass-fed animal meats, vegetables, fruits, and nuts and was devoid of grains, processed milk and milk products, refined sugar, salt, and seed oils. Caveman diet/hunter-gatherer's diet was popularised in mid-1970s by a gastroenterologist named Walter L. Voegtlin (1). The hunter-gatherer's diet was more for sustenance, whereas in the modern world, food is an (with due respect to the millions of poor who find it difficult to get one square meal per day) occasion for indulgence and gastronomic extravagance. The proponents of the caveman diet argue that this was the kind of diet followed by humans for millions of years, and therefore, we are genetically adapted to it. They argue the present diet that consists of grains and refined foods are not suitable for our genetic make-up, as it takes millions of years for genetic adaptations. It is true that we cannot go back to the olden days and live like cavemen, but there is some truth in their observation that our present diet is making people sick.

Initially, our knowledge about food was narrowed down to caloric values, macronutrients (carbs, protein, and fat), and micronutrients (vitamins and minerals). Factors like food complexity, form, response of body to different food (natural, altered, and synthetic) products were not given much importance. We all know sugar cane is not same as refined cane sugar nor is beetroot the same as refined beet sugar. The former is more complex with it is fibre, phytonutrients, vitamins, minerals, and natural colouring pigment. The latter is devoid of all the nutrients but packed with sweetness, calories, and extra chemicals!

In early civilisation, people depended on natural foods that were consumed mainly in the raw form, and with the advent of fire, we started cooking. Even then, we predominantly used a slow mode of cooking, unlike the express cooking methods available today! Food that is cooked over a slow fire is known to retain more nutrients compared to food cooked with a rapid modern cooking method. There are also differences in health benefits between dry and moist cooking.

Food is the source of all nutrients required for proper day-to-day functions of trillions of cells in our body. The quality and quantity of nutrients are important determinants for proper physiological functioning of the cells.

Inflammatory Foods

Inflammation is an important pathophysiological factor in many diseases, including cardiovascular, Alzheimer's, cancers, arthritis, and diabetes. Many researchers have identified inflammation as a potential mechanism of diabetes causation.

Half a century ago, the understanding of the relationship of diet and disease was restricted to a low-protein diet in kidney diseases, a low-salt diet in hypertension and nephrotic syndrome, a gluten-free diet in gluten sensitivity and coeliac diseases, a low phenylalanine diet for phenylketonuria, and a low-purines diet for gout.

The realisation that ordinary foods could cause inflammation and chronic diseases is relatively new and, until now, has not been widely acknowledged by mainstream medicine.

How do food, believed to be a friend helping to nourish our bodies, turn into a foe? In the coming sections, we will examine how diet can promote an inflammatory response.

Knowing how to detect chronic inflammation in the system (systemic inflammation) can also be helpful, as most inflammation leads later to chronic diseases. Some laboratory tests can detect inflammation in the system; these rely on the presence of inflammatory markers. The important inflammatory markers are C-reactive protein (CRP), fibrinogen, serum amyloid A protein, white cell count, ESR, albumin, cytokines (e.g., IL 1β, IL 6, TNF-α), and adhesion molecules (e.g., ICAM-1, VCAM-1). Many of these markers are used in research, and some are markers for acute inflammation (e.g., CRP, fibrinogen).

High-sensitivity C-reactive protein (HS-CRP) has been used as an inflammatory marker in various studies; for example, a Dutch study found that consumption of a high glycaemic index food increases the levels of HS-CRP. According to their findings, every ten-unit increase in dietary GI was associated with a 29% increase in HS-CRP (2). High-fibre intake has been shown to lower HS-CRP in some studies (3). A major clinical trial, the Women's Health Initiative, determined that a high-fibre diet is associated with low levels of interluekin-6 (IL-6) and tumour necrosis factor-alpha (TNF-α), which are both important mediators of inflammation (4).

In 1997, John C. Pickup first proposed the hypothesis that chronic low-grade inflammation and activation of innate immune system are closely associated with pathogenesis of type 2 diabetes (5). Acute phase reactants—which are circulating markers of inflammation—like interleukin (IL)-6 (a major cytokine mediator of the acute-phase response) are strong predictors of the development of type 2 diabetes (6).

In a recent article Pickup and his colleague, Fernandez, stress that certain components in the innate immune system in muscle, bone, liver, adipose tissue, and macrophages play a role in insulin action. According to them, exposure to acute infections (such as in pandemics) and chronic low exposure to environmental agents increase the susceptibility and surge in insulin resistance and type 2 diabetes (7). This point considered along with other factors gives us a reasonable explanation for the exponential increase of diabetes in the current population.

Dietary factors also appear to have an effect on inflammation and endothelial functions (cells which lines the blood vessels), irrespective of other factors, such as smoking, high cholesterol, high homocysteine, and high blood pressure (8).

A study conducted on nurses' health (Nurses' Health Study 1 and 2) and dietary patterns, markers of inflammation, and incidence of type 2 diabetes, it was determined that diets high in sugary foods, *Diet* soft drinks, refined grains and processed meat, low in vegetables (cruciferous and yellow vegetables), consuming wine and coffee might increase the risk of developing type 2 diabetes, possibly by aggravating inflammatory processes (9).

Maintaining an ideal weight and following a prudent diet (one that is plant-based with a low intake of red meat, meat products, sweets, high-fat dairy, and refined grains) or a Mediterranean diet (one rich in olive oil, fruits and vegetables, with whole grains, pulses, nuts, dairy products, and moderate red wine consumption) are the best strategies for decreasing diabetes risk (10). When choosing a diet, it is important to consider dietary preferences, as this might help people to stick on to a healthy diet on a longer basis.

Considering both these recommendations, I would suggest to aim for a diet that includes less processed dairy, less red and/or processed meat (and less heme-containing foods), and more whole grains (avoiding refined carbohydrates in the form of refined wheat flour—most of the white bread and bakery stuff is made up of this—and white rice), fish, pulses, nuts, and plenty of vegetables (especially raw), and reasonable quantities of fruits would be the best strategy for the prevention and control of diabetes and cardiovascular complications.

Fat is not the culprit: the type of fat is important

The traditional line of thought among mainstream medical practitioners is that all forms of dietary fat are bad and can cause obesity, high blood cholesterol, and coronary artery disease. Hence most of the medical practitioners insisted on low fat and low calorie diets. Food manufacturers also responded by bringing out products labelled low fat, lite, and fat-free. However, that shift promoted an overconsumption of highly refined carbohydrates. We now realised that these modifications could not bring down the number of obesity and diabetes. The occurrences of these disorders keep on increasing in many countries despite a reduction in the proportion of calories from fat. This observation in the USA has resulted in this phenomenon being named The American Paradox. The reasons mentioned for this paradox are the overriding effect of decreased physical activity, overconsumption of highly palatable, energy-dense, carbohydrate-rich, low-fat products, and under-reporting of fat consumed in dietary surveys (11,12). Evidences show that excessive intake of carbohydrates can lead on to increased fat deposition by suppressing lipolysis and fat oxidation.

According to Walter, the dietary fat is not the primary cause of excess body fat (at least in American society) and hence reduction of fat is not the solution (13). The type of fat is more important than the quantity of it. Evidences are against the theory of 'all saturated fats are bad and all polyunsaturated fats (PUFA) are good.' Western diet is predominantly constituted of omega-6 fatty acid—which is a PUFA. The high ratio of omega-6 to omega-3 (10:1) in these diets enhances production of circulating and cellular pro-inflammatory mediators and reduces anti-inflammatory mediators. The resultant overall increase in systemic inflammation is hypothesised as the reason for the higher incidence of allergic and inflammatory diseases including asthma, allergies, diabetes, cardiovascular disease, and arthritis (14).

Long-chain FFAs induce several cyto—and chemokines in human islets. The FFAs most abundantly found in human nutrition-oleate, palmitate, and stearate—stimulate IL-1β expression (15). The stimulatory effects of FFAs on pro-inflammatory mediators are observed in islet cells as well as in numerous other cell types, such as muscle, macrophage, adipocyte cell lines, and in coronary artery endothelial cells (15). The combination of FFAs with elevated glucose becomes a double whammy for the generation of inflammatory mediators. Daily supplementation of 1-8 g of omega-3 PUFA in the diet has shown to reduce inflammation and found to be effective in the treatment of IBD, eczema, psoriasis, ulcerative colitis, and rheumatoid arthritis (16).

Curtis et al. have shown that linolenic acid, EPA, or DHA reduce the expression of genes for TNF and IL-1 in chondrocytes (bovine). Thus supplementation with omega 3 fatty acids can modulate the expression and activity of inflammatory and degradative factors responsible for cartilage destruction in arthritis (14,17).

Recent studies have shown that diets with palmitic, oleic, lauric, and myristic acids do not alter inflammatory markers (homocysteine) in healthy adults (18).

Lifestyle and Diabetes Risk

Five lifestyle intervention studies demonstrated that combination of lifestyle modifications—increased physical activity, diet, and weight reduction—delay the development of type 2 diabetes in individuals with impaired glucose tolerance. They are Malmö study from Sweden; Da Qing Study from China; The Finnish Diabetes Prevention Study (DPS); The Diabetes Prevention Program (DPP) in USA, and The Indian Diabetes Prevention Study (19).

Patients were allotted into three groups in the Da Qing study. The first group was treated with exercise intervention, the second with diet, and the third combined. They found the incidence of type 2 diabetes during six-year follow-up was low (41%) in exercise group compared with diet alone (44%) or combination group (46%).

Similar result was observed in the DPS study. As per their findings, people who increased their physical activity were 66% less likely to develop type 2 diabetes. In DPP, people who met the activity goals had 44% reduction in diabetes incidence.

Does food affect our genes?

In both DPP and DPS, the risk of developing type 2 diabetes was reduced by 58% over three years period. In DPS, the risk of developing type 2 diabetes was high in those with genetic polymorphism. Among the fourteen genes identified, two (PPARG2 and TCF7L2) are confirmed to increase the risk of developing type 2 diabetes in large case-control studies. All these are evidences for a gene—lifestyle interaction.

Chances of getting diabetes were higher in people with TDF2L2 genotype in DPP study but found to be reduced in both lifestyle and metformin interventional groups (20).

TCF7L2 is the strongest type 2 diabetes locus identified so far (21). It plays an important role in insulin synthesis, processing, and secretion. Foods that increase insulin demand might increase the risk of getting type 2 diabetes in TCF7L2 variants. The quality and quantity of carbohydrates can also modify the risk of diabetes in these groups of patients.

Epigenetic Changes

Epigenetics is the study of heritable changes in the gene function that occur without a change in DNA sequence.

Epigenetic mechanisms are implicated in gene regulation and development of different diseases. The epigenome differs between cell types and can be altered by environmental factors.

The main mechanisms through which epigenetic changes occur are DNA methylation and chromatin remodelling. Many nutrients in food are found to have an effect in DNA methylation. Deficiencies of choline, methionine, folate, vitamin B 12, vitamin B6, and riboflavin affect one-carbon metabolism. S-adenosyl methionine (SAMe) is generated through one-carbon metabolism. SAMe is an important component in methylation process. This results in impaired DNA methylation and increase the risk of neural tube defects, cancer, and cardiovascular diseases (22).

Chromatin remodelling (alters accessibility of DNA for transcription) is partly regulated by the energy balance inside the cell. Change in calorie intake is shown to alter chromatin remodelling. Diet or environmental factors altering the level of protein, enzymes, and RNA involved in the chromatin remodelling might also be considered as a mechanism of regulating the gene expression.

Long-term exposure of diets that influence chromatin remodelling and DNA methylation could induce permanent epigenetic changes in the genome.

Single Nucleotide Polymorphism/SNPs

Snips are DNA sequence variations that occur when a single nucleotide (the chemicals called as bases and represented by alphabets A, C, T, G) in the genome sequence is altered. The genome sequence AAGGCTA can be changed to ATGGCTA in SNPs. The SNP is considered only if it occurs at least in 1% of the population. SNPs constitute about 90% of human genetic variation.

Nutritional Genomics

This branch of science studies how foods affect gene expression (nutrigenomics) and how genetic variation in individual affects his response to a particular food (nutrigenetics).

The basic tenets of the nutritional genomics are the following:

- Dietary chemicals can alter gene expression or structure.
- Diet interacts with genes and environmental factors to produce serious risk factor in number of diseases.
- The degree of influence on health and disease depends on individual genetic make-up.
- Genes regulated by dietary chemicals likely to play protection against chronic diseases or delaying its onset.

GI and GL

Glycaemic index (GI) gives an idea of how quickly blood sugar levels are raised after eating a particular food. It is a relative index comparing the raise in blood glucose level after consuming a food with pure glucose or white bread. The main drawback of this index is that it does not consider the actual quantity of carbohydrate consumed.

Glycaemic Load is developed to overcome this shortcoming. Glycaemic load (GL) takes into account the quantity of carbohydrate present in the food and how much each gram of it can raise the blood glucose level. It is based on glycaemic index and is defined as the grams of available carbohydrate (available carbohydrate = total carbohydrate minus the fibre) in the food times the food's GI divided by 100.

Foods are classified into three groups as per GI. Low is food with values 55 and less, medium is between 56 and 69, high is equal to or above 70.

For GL, the values are, 10 for low, 10-19 for moderate, and >20 for high.

Low GI foods release glucose slowly and steadily, whereas the high GI are like rocket fuel! Some examples of high GI foods are white bread, glucose, and white rice. The rule of the thumb is, the more fibre you are having in a particular food, the less would be its GI value.

The GL is preferable than GI as it considers the quantity of carbohydrate in the food. This can be easily explained in the case of watermelon. The GI of watermelon is 72, and it comes in the category of high GI food. We all know that majority of its weight is attributed to its water content. A 100 g serving of watermelon has 5 g of carbohydrate in it. So if we calculate the GL, ($72 \times 5/100 = 3.6$), it is low.

The epidemiological studies suggest that consuming high GI foods is an independent risk factor for developing diabetes (23). Moreover the degree of satiety is also related to GI of the carbohydrates in the food (24).

Dietary fibres are resistant to digestive enzymes and prolong the satiety. By modulating satiety signals, dietary fibres can induce early satiety. When combined with high carbohydrate, high palatable foods, they can reduce the total energy intake by minimising consumption of high-calorie foods (25).

- Jump from cavemen diet to a modern diet is believed to be one of the reasons for the recent surge in chronic inflammatory disorders.
- Food is not inert but has a potential role in our immune system, generation of inflammation, genetic changes, and in lot of other unknown influences in our body.
- Following a healthy diet is important in preventing and reversing many of the lifestyle induced disorders.
- Natural foods in low GI, high fibre and with sufficient nutrients are important in proper metabolic functions of the cells.
- Food and lifestyle are potent enough to make changes in our genes.

References

1. http://en.wikipedia.org/wiki/Paleolithic_diet
2. Du H, van der A DL, van Bakel MM, van der Kallen CJ, Blaak EE, van Greevenbroek MM, Jansen EH, Nijpels G, Stehouwer CD, Dekker JM, Feskens EJ. Glycemic index and glycemic load in relation to food and nutrient intake and metabolic risk factors in a Dutch population. Am J Clin Nutr. 2008;87(3):655-61
3. North CJ, Venter CS, Jerling JC. The effects of dietary fibre on C-reactive protein: an inflammation marker predicting cardiovascular disease. Eur J Clin Nutr. 2009;63:921-33

4. Ma Y, Hébert JR, Li W, Bertone-Johnson ER, Olendzki B, Pagoto SL, Tinker L, Rosal MC, Ockene IS, Ockene JK, Griffith JA, Liu S. Association between dietary fiber and markers of systemic inflammation in the Women's Health Initiative Observational Study. Nutrition. 2008;24(10):941-9

5. Pickup JC, Mattock MB, Chusney GD, Burt D. NIDDM as a disease of the innate immune system: association of acute-phase reactants and interleukin-6 with metabolic syndrome X. Diabetologica. 1997 Nov;40(11):1286-92; PMID 9389420

6. Pickup JC. Inflammation and activated innate immunity in the pathogenesis of type 2 diabetes. Diabetes Care. 2004;27:813-23; PMID: 14988310

7. Fernández-Real JM, Pickup JC. Innate immunity, insulin resistance and type 2 diabetes. Diabetologia. 2012;55(2):273-8. PMID: 22124608

8. Brown AA, Hu FB. Dietary modulation of endothelial function: implications for cardiovascular disease. Am J Clin Nutr. 2001;73:673-86. PMID: 11273841

9. Schulze MB. Dietary pattern, inflammation and incidence of type 2 diabetes in women. Am J Clin Nutr. 2005;82(3):675-715. PMID: PMC2563043

10. Salas-Salvadó J. The role of diet in the prevention of type 2 diabetes. Nutr Metab Cardiovasc Dis. 2011;21(Suppl 2):B32-48. PMID: 21745730

11. Astrup A, Lundsgaard C, Stock MJ. Is obesity contagious? Int J Obes. 1998;22:375-6

12. Astrup A. The American paradox: the role of energy-dense fat-reduced food in the increasing

prevalence of obesity. Curr Opin Clin Nutr Metab Care. 1998;1(6):573-7; PMID: 10565412

13. Willett WC, Leibel RL. Dietary fat is not a major determinant of body fat. Am J Med. 2002;113(9 Suppl 2):47-59

14. Weaver KL, Ivester P, Seeds M, Case LD, Arm JP, Chilton FH. Effect of dietary fatty acids on inflammatory gene expression in healthy humans. J Biol Chem. 2009;284(23):15400-7

15. Donath MY, Böni-Schnetzler M. Islet inflammation impairs the pancreatic β-cell in type 2 diabetes. Physiology. 2009;24:325-31; PMID: 19996363

16. Gil Á. Polyunsaturated fatty acids and inflammatory diseases. Biomed Pharmacother. 2002;56:388-96

17. Curtis CL, Hughes CE, Flannery CR, Little CB, Harwood JL, Caterson B. n-3 Fatty acids specifically modulate catabolic factors involved in articular cartilage degradation. J Biol Chem. 2000;275:721-4

18. Voon PT, Ng TK, Lee VK, Nesaretnam K. Diets high in palmitic acid (16:0), lauric and myristic acids (12:0 + 14:0), or oleic acid (18:1) do not alter postprandial or fasting plasma homocysteine and inflammatory markers in healthy Malaysian adults. Am J Clin Nutr. 2011;94(6):1451-7; PMID: 22030224

19. Kilpeläinen T. Physical activity, genetic variation, and type 2 diabetes. Kuopio University Publications. D. Medical Sciences. 2009; 462

20. Jose C. Florez, Kathleen A. Jablonski, Nick Bayley, Toni I. Pollin, Paul I.W. de Bakker, Alan R. Shuldiner, William C. Knowler,David M. Nathan, David Altshuler. TCF7L2 polymorphisms and progression to

diabetes in the diabetes prevention program. N Engl J Med. 2006;355(3):241-50

21. Cornelis MC, Qi L, Kraft P, Hu FB. TCF7L2, dietary carbohydrate, and risk of type 2 diabetes in US women. Am J Clin Nutr. 2009;89(4):1256-62

22. Stover PJ, Garza C. Bringing individuality to public health recommendations. J Nutr 2002;132:2476s—80s

23. Salmeron J, Manson JE, Stampfer MJ, Colditz GA, Wing AL, Willett WC. Dietary fiber, glycemic load, and risk of noninsulin-dependent diabetes mellitus in women. JAMA. 1997;277:472-7; PMID: 9020271

24. Ludwig DS, Majzoub JA, Al-Zahrani A, Dallal GE, Blanco I, Roberts SB. High glycemic index foods, overeating, and obesity. Pediatrics. 1999;103:26

25. Salmeron J, Ascherio A, Rimm EB, et al. Dietary fiber, glycemic load, and risk of NIDDM in men. Diabetes Care. 1997;20(4):545-50; PMID:9096978

CHAPTER 7

Power of Exercises

Just ask yourself these questions:
> Are you doing any physical activities?
> How often you do it?
> How much are you doing?

Experts working in the field of obesity and diabetes agree that activity levels of people have reduced significantly in the past decades. The contributing factors are many—industrialisation, internet, declining need of physical exertion at job sites, and better transport facilities. In addition, the rapid progress in the technology makes people sit for longer hours—spending more time on computer, television, virtual games instead of real games, sitting in car, and at office desk. Our home and office environments are made more easy and sedentary by gadgets and computers. In majority of the urban societies, physical activity has to be a planned and purposeful one than an automatic one. These are not sufficient enough to offset the extra energy gained through consuming high-calorie foods and energy saved through reducing the activity levels. In an article titled 'Obesity in Britain: Gluttony or Sloth', authors Prentice and Jebb mention that hours per week of television viewing and number of cars per household are closely related to the current trends of obesity epidemic in Britain (1). I think this observation is true in the case of most developed societies. Various studies—prospective as well as clinical trials—show that moderate or high level of physical activity or physical fitness and changes in lifestyle can prevent type 2 diabetes (2). Regular physical activity can reduce death from all causes by protecting against atherosclerosis (hardening of arteries), type 2 diabetes, colon cancer, and breast cancer.

The topic about exercise would not be complete without mentioning thermodynamics. It deals with heat and its relationship with energy and work. It is best applicable to mechanical systems than to biological systems like human body. The law of thermodynamics and thermodynamic equation are often quoted in the context of weight loss.

The first law of thermodynamics mentions energy in a closed system is neither created nor destroyed (3). But human body is not a machine, and it is not a closed system. So there are limitations for its application in a biological system. It is a widely held belief that energy going in should be equated by energy burned out to maintain a balanced body weight. The following tenets are considered in this approach.

- Calorie counting.
- One pound of fat = 3,500 kcal/1 kg Fat = 7,700 kcal.
- If you are taking 100 kcal of food, you must do an exercise which burns 100 kcal to maintain your weight.

So people tend to believe a Snicker's bar (15 g) which gives around 71 kcal is better than a Gala apple as it is 75 kcal!

Exercise produces anti-inflammatory effect

Our body produces substances enhancing and inhibiting inflammation. During the course of an infection, the inflammatory cytokines like IL1β and TNF α are released into the circulation. There is release of cytokines during exercise but not up to the same level as during infection. The classic pro-inflammatory cytokines (IL1β and TNF-α) are not released during exercise, whereas the level of another cytokine (IL-6) can be raised up to 100 folds. This returns to base line after exercise. The exercise intensity, duration, the mass of muscle recruited and the person's endurance capacity are determinants of raised blood levels of IL-6 (Pederson) (4). The IL-6 can be either pro—or anti-inflammatory. But studies have shown IL-6 has more anti-inflammatory action than producing it. IL-6 reduces the release of inflammatory cytokines like TNFα and IL-1 and enhances the release of anti-inflammatory cytokines IL-10, IL-1ra (IL-1 receptor antagonist). In addition to it, exercise releases anti-inflammatory cytokines or cytokine inhibitors like IL1ra and sTNF-R (5).

The genetic polymorphism is an important factor to be considered in both obesity and diabetes causation. In most of the cases, just having the causative gene is not enough to produce disease. The interaction of environmental factors may be crucial for their expression. Polymorphism of the β-2-adrenoceptor gene was shown to be associated with obesity in a Swedish study (6). A French study demonstrated the presence of polymorphism failed to produce obesity on physically active people but was associated with an excess of body weight (7 kg) and larger waist circumference in those without physical activity (7).

Exercise, Fat, and Insulin Resistance

The fat accumulated in the abdomen is known as visceral fat. Excessive visceral fat increases insulin resistance and leads on to type 2 diabetes. Regular physical activity reduces the risk of diabetes and obesity by reducing the total and visceral fat. Exercise training increases insulin sensitivity of skeletal muscles, increases muscle content of glycogen synthase (GS), and glucose transporter isoform 4 (GLUT4). Increase in GS accelerates the non-oxidative disposal of glucose as glycogen. GLUT4 increases the entry of glucose into the muscle cells thereby reducing circulating blood glucose. Exercise also increases the oxidative capacity of skeletal muscles, which results in higher levels fat oxidation at rest as well as during submaximal exercises. This prevents the lipid-mediated insulin resistance. Studies done in sedentary people have demonstrated that the accumulated triglycerides in muscles increases insulin resistance. Metabolites of triglycerides such as fatty acyl-CoA, diacylglycerols, and ceramides may contribute to insulin resistance. Increased lipid oxidation induced by exercise improves fatty acid turnover. This prevents accumulation of fatty acids metabolites in the muscles, which in turn leads on to improved insulin sensitivity (8).

Regular exercise increases the microvasculature to muscles; increase in blood flow increases the uptake of insulin and glucose into the muscles. We have already seen the role of low-grade systemic inflammation as a contributing factor for many diseases, including diabetes. Regular physical activity reduces inflammation by increasing the anti-inflammatory cytokines released during exercise. There is about 100-fold increase in IL-6 after exercise. Even though IL-6 is a pro-inflammatory in nature, it stimulates the secretion of other anti-inflammatory cytokines like IL-1 receptor antagonists, soluble TNF α receptor, and IL-10.

Genes and SNPs

Cell is the smallest functional unit in a biological system. The nucleus, the central controlling unit of the cell, has DNA which carries our genetic material. We are all born with a set of genes which cannot be changed. The methyl groups attached to our DNA can activate or deactivate a gene. The methyl group can be influenced by exercise, diet, and lifestyle through a process called as DNA methylation. Changes in DNA methylation have been suggested to be a biological mechanism behind the beneficial effects of physical activity.

A recently published study shows the impact of long-term exercise in DNA methylation in human skeletal muscle. They found out that exercise for six months is associated with epigenetic changes (9). These epigenetic alterations are important for preventing type 2 DM in patients with positive family history (9).

In another six months intervention study, researchers identified the power of exercise to influence the genome-wide DNA methylation pattern in human adipose tissues. This proves us beyond doubt that exercise is a powerful factor which can modify DNA methylation (10).

As we have already seen that food and environment interacting with SNPs, there are indications of a possible association of SNPs with exercises.

Two identified SNPs in adiponectin gene or ACDC gene (rs2241766 and rs1501299) can affect the levels of circulating adiponectins, insulin resistance, and type 2 diabetes.

Researchers, when studied the association of SNPs and metabolic risk factors among 1600 Korean adults, found reasonable evidences to speculate that metabolic environments (poor cardio respiratory fitness and insulin resistance) could affect the regulation of the ACDC gene and the SNP within it (11).

Exercise and Telomere length

Telomeres are end portions of chromosomes and composed of non-coding DNA sequence. The length of telomere gets shorter with each cell cycle, means as we age, the telomere becomes shorter. Short telomeres are associated with increased risk of age-related problems. Older endurance trained athletes found to have longer telomere length compared with people in the same age group doing moderate exercises. In a study comparing two groups of people with telomere length and exercise capacity (VO2 max), people with higher VO2 max have found to have longer telomere length (12).

Lifestyle intervention by counselling for physical activity, healthy eating habits, and body weight can reduce the risk of diabetes by 40-60% among adults with impaired glucose tolerance (13).

- Calorie counting is not the best way of approaching healthy eating.
- Overindulging in unhealthy foods and trying to offset it through exercise would probably not work in the long run.
- Exercise is powerful enough to reduce inflammation, insulin resistance, change in SNPs and telomere length.
- Lifestyle intervention can reduce risk of diabetes by 40-60% in adults having impaired blood sugar.

References

1. Prentice AM, Jebb SA. Obesity in Britain: gluttony or sloth? Br Med J. 1995;311:437-9; PMID: 7640595 PMCID: PMC2550498

2. Hu G. Exercise, genetics and prevention of type 2 diabetes. Essays Biochem. 2006;42:177-92; PMID: 17144888

3. http://en.wikipedia.org/wiki/First_law_of_ thermodynamics

4. Pedersen BK, Steensberg A, Schjerling P. Muscle-derived interleukin-6: possible biological effects. J Physiol. 2001;536:329-37

5. Petersen AM, Pedersen BK. The anti-inflammatory effect of exercise. J Appl Physiol. 2005;98:1154-62; PMID: 15772055

6. Meirhaeghe A, Helbecque N, Cottel D, Amouyel P. β 2-adrenoceptor gene polymorphism, body weight and physical activity. Lancet. 1999;353:896; PMID: 10093985

7. Astrup A, Raben A, Buemann B, Toubro S. Fat metabolism in the predisposition to obesity. Ann N Y Acad Sci. 1997;827:427-33

8. Kilpeläinen T. Physical activity, genetic variation, and type 2 diabetes. Kuopio University Publications. D. Medical Sciences. 2009; 462

9. Nitert MD, Dayeh T, Volkov P, Elgzyri T, Hall E, et al. Impact of an exercise intervention on DNA methylation in skeletal muscle from first-degree relatives of patients with type 2 diabetes. Diabetes. 2012; 61(12):3322-32.

10. Rönn T, Volkov P, Davegårdh C, Dayeh T, Hall E, et al. A six months exercise intervention influences the genome-wide DNA methylation pattern in human adipose tissue. PLoS Genet. 2013; 9(6):e1003572.

11. Lee JY, Cho JK, Hong HR, Jin YY, Kang HS.Genetic effects of adiponectin single nucleotide polymorphisms on the clustering of metabolic risk factors in young Korean adults. Eur J Appl Physiol. 2012; 112(2):623-9.

12. Ida Beate Ø. Østhus, Antonella Sgura, Francesco Berardinelli, Ingvild Vatten Alsnes, Eivind Brønstad, Tommy Rehn, Per Kristian Støbakk, Ha°vard Hatle, Ulrik Wisløff, Javaid Nauman. Telomere length and long-term endurance exercise: does exercise training affect biological age? A pilot study. PLoS One. 2012;7(12):e 52769

13. Qi L, Hu FB, Hu G. Genes, environment, and interactions in prevention of type 2 diabetes: a focus on physical activity and lifestyle changes. Curr Mol Med. 2008;8(6):519-32; PMID: 18781959

CHAPTER 8

Nutraceuticals and Supplements

These groups of products are not 'star performers' like some of the well-known drugs in conventional medicine. Yet a lot of people are using them either for controlling their blood sugar or for reducing complications. Sometimes they are even prescribed by the modern medicine practitioners when they find it difficult to attain glycaemic control in patients on conventional medications.

For some of the main-stream medicine practitioners (Western medicine practitioners) and researchers, they are *shaman* in nature, but there are supporting evidences for their efficacy. Let us have a look into some of the common neutraceuticals/supplements used for diabetes. I classified them into two broad categories: (1) Vitamins and Minerals and (2) Herbs.

Vitamins and Minerals

We have seen that simple sugars and refined food products coupled with low physical activity are the main reasons for diabetic epidemics. The refined foods are low in fibre, phytonutrients, minerals, and vitamins. Minerals like chromium, magnesium, zinc, vanadium, and manganese are found to be critical in diabetic population. Among this, chromium is the most limiting (in the diet) and having significant effects on diabetes.

Chromium

Chromium (Cr) is an essential nutrient involved in metabolism of glucose, insulin, and blood lipids. It is a trace element found in different tissues in human body. Chromium is thought to be a cofactor necessary for optimal insulin action. Therefore, chromium supplementation may exert its potential benefits by improving insulin sensitivity. Chromium plus two molecules of nicotinic acid forms a biological compound called as Glucose Tolerance Factor, which has got properties of enhancing the action of insulin. Cr also increases the number of insulin receptors and activates insulin receptor kinase leading to increased insulin sensitivity.

Chromium deficiency is associated with lipid abnormalities and increased risk of atherosclerosis. National Research Council of National Academy of Science recommends Cr intake of 50-200 µg/day. Standard North American diet has <50 µg of Cr. Moris et al. showed that Cr levels are low in diabetics compared with non-diabetic people (1). They also found that low Cr levels are related to elevated plasma glucose concentrations (2). Cr is widely distributed in human body, and there are no established standards to estimate total body Cr stores.

Lowering of serum triglycerides was noted when Chromium picolinate at a dosage of 200 µg/day given for two months in diabetic people (3). Higher dosages of Cr (1,000 µg/day) have been found to have better effects than lower dosages (4). These researchers found that improvements in blood sugar and insulin response normally occur within a few weeks after starting Cr, but it took a longer duration for improvements in blood lipids levels.

Trivalent Cr is the form used in foods and supplements, and it is one of the least toxic nutrients. According to USEPA, the reference dosage is 350 times than the advised daily dietary intake, which is 200 µg.

A meta-analysis done on forty-one studies also concluded that Cr supplementation in patients with type 2 diabetes may have a modest benefit on glycaemia and dyslipidemia (5).

Zinc

The relationship between zinc and insulin was known to man as early as 1930s, when commercial insulin was produced in the form of protamine zinc insulin and lente insulin. In these preparations, zinc was added to prolong the duration of action of insulin. Insulin is secreted in the β cells of pancreas as a single chain peptide which is doubled up and linked together with a double sulfur bond (inter chain disulfide bond). This is known as proinsulin, and it is broken to remove the C-peptide to make it two chains (alpha and beta peptide chains). These chains are known as insulin monomers and made up of fifty-one amino acids each. They are linked together by disufide bonds.

Zinc is important in insulin action and carbohydrate metabolism. Zinc is a structural part of many key antioxidant enzymes such as superoxide dismutase, and zinc deficiency impairs their synthesis, leading to increased oxidative stress.

In the presence of zinc inside the cell, the insulin is assembled for storage and secretion as zinc crystal. Through video fluorescence analysis, Zalewski demonstrated that concentration of zinc inside the islet cell is related to synthesis, storage, and secretion of Insulin (6).

It is found that serum levels of zinc were 40% low in diabetic people compared with normal. Zinc excretion is found to be high in urine of diabetic people. It is thought that hyperglycaemia is responsible for this finding as it interferes with the active transport of zinc back into the renal tubular cells. In a study done in 175 diabetic patients, positive correlation was noted between urinary zinc excretion and HbA1c levels (7). It is also found that hyperzincuria is a universal finding in diabetic people.

As I have already mentioned in previous sections, higher levels of insulin are secreted in early stages of type 2 diabetes. As zinc leaves the cell with insulin, this results in low levels of zinc inside the cell. This is also compounded by the hyperzincuria found in diabetics. Later on, with less available zinc, less insulin is secreted for a given glucose level, which then can lead on to islet cell damage.

In a RCT study, ninety-six diabetic people were divided into three groups: first group with oral zinc+ multivitamin mineral (MVM); second one with MVM alone, and third one with placebo for four months. At the end of the study period, there was significant reduction in fasting blood sugar and HbA1c noted in the group on Zn+MVM (8).

Another randomised double-blind placebo study done in Tunisian diabetic population comparing zinc (30 mg/day of zinc gluconate for six months) with placebo found that plasma thiobarbituric acid reactive substances (TBARS) was decreased 15% in treated group. TBARS is a marker of lipid peroxidation, and reduction of it indicates antioxidant potential of Zn (9).

In a systemic review of meta-analysis Jayawardena et al. found that zinc supplementation improves glycaemic control, lipid parameters, antioxidant status, and reduction in blood pressure (10).

Magnesium

Magnesium (Mg) is an important cofactor for enzymes of carbohydrate metabolism. Magnesium concentration inside the cells (intracellular magnesium) is critical for several enzymes (especially those involved in phosphorylation reactions such as tyrosine-kinase) in carbohydrate metabolism. There is a strong relationship between Mg and insulin resistance (11).

There are several reports indicating the presence of low intracellular free magnesium concentrations in subjects with essential hypertension or type 2 diabetes. Intracellular free magnesium deficiency may be an important link between insulin resistance, hypertension, and accelerated cardiovascular disease (12).

Magnesium is mainly seen in unprocessed foods like whole grain, nuts, green-leafy vegetables and is lost during processing.

In Atherosclerosis Risk in Communities Study (ARIC) low blood levels of magnesium was found to be common among diabetic people.

In a study including two large prospective cohorts (group of people enrolled and monitored for a period) provide strong and consistent evidence for an inverse association between magnesium intake and diabetes risk (13).

Vanadium

It is a trace element found in human tissues and serves as essential cofactor in various enzymatic reactions. Studies show that vanadium mimics the action of insulin (14). Vanadium might be having potential beneficial effects in diabetic people. Dietary sources for vanadium include mushrooms, shellfish, black pepper, parsley, dill seed, and grains.

Vanadium in the form of vanadyl sulfate is shown to regulate blood sugar levels and improve insulin sensitivity. Researches show beneficial effects of vanadium in type 2 diabetes with insulin resistance.

Improvements in blood sugar and insulin sensitivity were found in vanadyl sulfate supplementation of 100 mg/day (15). Higher dosage, up to 300 mg/day, showed an improved glycaemic control (16).

Vanadium is having toxicity, but Vanadyl sulfate is 6-10 times less toxic than vanadate. Organic and chelated ions of vanadium are less toxic compared with inorganic vanadium salts.

Vitamin D

Recent evidences show that vitamin D has more important roles in human health than thought initially, preventing rickets and osteomalacia (the condition of bone softening). It has immunological and blood sugar regulatory properties.

Its role in type 1 diabetes is widely reported and newer studies show that it is important in type 2 diabetes too. There are reports that vitamin D deficiency may predispose to glucose intolerance, altered insulin secretion, and type 2 diabetes.

Data from several studies have shown that hypovitaminosis D (low vitamin D) might play an important role in the pathogenesis of type 2 diabetes in human beings.

Vitamin deficiency has shown to alter insulin synthesis and secretion both in humans and animal models. Vitamin D deficiency predisposes to glucose intolerance, altered insulin secretion, and type 2 diabetes. Its role in type 2 diabetes was established when diabetics with low vitamin D improved blood glucose and insulin secretion by correcting their vitamin D levels (17).

Vitamin D receptors (VDR) and vitamin D-binding proteins (DBP) are identified in pancreatic tissues. Relationship between certain alleles of VDR and DBP genes with glucose tolerance and insulin secretion supports the role of vitamin D in type 2 diabetes. The mechanism of vitamin D deficiency in type 2 diabetes is thought to be mediated through regulation of plasma calcium (which in turn regulate insulin synthesis and secretion) and through its direct action on pancreatic β cells (17).

The proposed mechanisms for vitamin D deficiency in diabetes are the following (18):

1. Genetic predisposition as in type 1 diabetes
2. Increased body mass index (BMI)
3. Albuminuria in type 1 and 2 diabetes

4. Increased renal excretion of vitamin D-binding protein (DBP) in type 1 and 2 diabetes

Others findings worth considering are (19) as follows:

1. Higher incidence of hypovitaminosis in type 1 and type 2 diabetes
2. Vitamin D receptors and vitamin D-binding protein found in pancreatic β cells and other immune cells
3. Allelic variation in genes involved in vitamin D metabolism and VDR are associated with glucose intolerance, insulin secretion and sensitivity, and inflammation
4. Pharmacological dosage of vitamin D3 preventing insulitis and type 1 diabetes in non-obese diabetic mice and other models
5. Increased insulin sensitivity and reduced inflammation in type 2 diabetes

Serum 25(OH) D3 level is a better indicator of vitamin D status than 1,25(OH)2 D3, as 1,25(OH)2 D3 is cleared rapidly from the body.

Apart from the bone and muscle tissues, many other vitamin D targets have been reported, such as heart, stomach, liver, brain, skin, pancreatic islets (β cells), thyroid, parathyroid, and adrenal glands and immune cells (17).

There is high prevalence of type 2 diabetes in obese individuals, and obesity is often associated with hypovitaminosis D. Being a fat soluble vitamin, vitamin D is efficiently deposited in body fat stores where it is no longer bioavailable. This could be an explanation for low serum vitamin D levels in obese.

- Chromium, zinc, magnesium, vanadium, and vit D are found to be important in regulating blood sugar.
- Natural foods rich in minerals are important in diabetics.
- Physical activity in outdoor environment can influence the vit D levels in blood, which in turn improve insulin function.

References

1. Morris BW, Kemp GJ, Hardisty CA. Plasma chromium and chromium excretion in diabetes (Letter). Clin Chem. 1985;31:334-5
2. Morris BW, Griffiths H, Kemp GJ. Correlations between abnormalities in chromium and glucose metabolism in a group of diabetics (Letter). Clin Chem. 1988;34:1525-6
3. Lee NA, MD, Reasner CA. Beneficial effect of chromium supplementation on serum triglyceride levels in NIDDM. Diabetes Care. 1991;17(12):1449-52

4. Anderson RA, Cheng N, Bryden NA, Polansky MM, Chi J, Feng J. Elevated intakes of supplemental chromium improve glucose and insulin variables of people with type II diabetes. Diabetes. 1997;46:1786-91

5. Ethan M. Balk, Alice H. Lichtenstein, Anastassios G. Pittas. Effect of chromium supplementation on glucose metabolism and lipids: a systematic review of randomized controlled trials. Diabetes Care. 2007;30:2154-63

6. Zalewski P, Millard S, Forbes I, Kapaniris O, Slavotinek S, Betts W, Ward A, LincolnS, Mahadevan I. Video image analysis of labile Zn in viable pancreatic islet cells using specific fluorescent probe for Zn. J Histochem Cytochem. 1994;42:877-84

7. Chausmer AB. Zinc, insulin and diabetes. J Am Coll Nutr. 1998;17(2):109-15

8. Gunasekara P, Hettiarachchi M, Liyanage C, Lekamwasam S. Blood sugar lowering effect of zinc and multi vitamin/mineral supplementation is dependent on initial fasting blood glucose. J Diabetol. 2011;1:2

9. Roussel AM, Kerkeni A, Zouari N, Mahjoub S, Matheau JM, Anderson RA. Antioxidant effects of zinc supplementation in Tunisians with type 2 diabetes mellitus. J Am Coll Nutr. 2003;22(4):316-21

10. Jayawardena R, Ranasinghe P, Galappatthy P, Malkanthi RLDK, Constantine GR, Katulanda P. Effects of zinc supplementation on diabetes mellitus: a systematic review and meta-analysis. Diabetol Metab Syndrome. 2012;4:13

11. Huertha MG, Roemmich JN, Kington ML, Bovbjerg VE, Weltman AL, Holmes VF, Patrie JT, Rogol AD, Nadler JL. Magnesium deficiency is associated with insulin resistance in obese children. Diabetes Care. 2005;28(5): 1175-81

12. Nadler JL, Buchanan T, Natarajan R, Antonipillai I, Bergman R, Rude R. Magnesium deficiency produces insulin resistance and increased thromboxane synthesis. Hypertension. 1993;21:1024-9

13. Lopez-Ridaura R, Willett W C, Rimm E B, Liu S, Manson J E, Stampfer M J, Hu F B. Magnesium intake and risk of type 2 diabetes in men and women. Diabetes Care. 2004;27:134-40

14. Poucheret P, Verma S, Grynpas MD, McNeill JH. Vanadium and diabetes. Mol Cell Biochem. 1998;188:73-80

15. Halberstam M, Cohen N, Shlimovich P, Rossetti L, Shamoon H. Oral vanadyl sulfate improves insulin sensitivity in NIDDM but not in obese nondiabetic subjects. Diabetes. 1996;45:659-66

16. Goldfine AB, Patti ME, Zuberi L, Goldstein BJ, LeBlanc R, Landaker EJ, Jiang ZY, Willsky GR, Kahn CR. Metabolic effects of vanadyl sulfate in humans with non-insulindependent diabetes mellitus: in vivo and in vitro studies. Metabolism. 2000;49(3):400-10

17. Palomer X, González-Clemente JM, Blanco-Vaca F, Mauricio D. Role of vitamin D in the pathogenesis of type 2 diabetes mellitus. Diabetes Obes Metab. 2008;10(3):185-97

18. Thrailkill KM, Fowlkes JL. The role of vitamin D in the metabolic homeostasis of diabetic bone. Clin Rev Bone Miner Metab. 2013;11(1):28-37

19. Takiishi T, Gysemans C, Bouillon R, Mathieu C. Vitamin D and diabetes. Endocrinol Metab Clin North Am. 2010;39(2):419-46

CHAPTER 9

Useful Herbs

Botanical products

Traditional medicines derived from plants are used by about 60% of the world's population. Many conventional drugs have been developed from medicinal plants. In the case of oral hypoglycaemic drugs, metformin is one such example. The plant *Galega officinalis* is used in traditional medicine for treating diabetes. It is rich in a hypoglycaemic compound guanidine which is toxic to humans. Later on, less toxic and related compounds were identified from this plant for treatment. The group of drugs named biguanides were later developed from this knowledge, and one of them is metformin. Metformin is one of the widely used prescription medication for diabetes in modern medicine.

Until now, around 400 traditional plants were reported for treatment in diabetes. But only a small number among them are scientifically evaluated. The hypoglycaemic effect of some herbal extracts has been confirmed in human and animal models of type 2 diabetes. The World Health Organization Expert Committee on diabetes has recommended that the role of herbs has to be investigated further (1,2).

The actions of the most of the botanical products are not restricted to blood sugar regulation. They also proved to be having effect on lipid metabolism, acting as antioxidant, and having favourable effects on blood pressure. Since they have multiple beneficial effects and in-depth analysis of all of them is beyond the scope of this book, I mention some of the most commonly used herbs.

Plenty of herbs are used in Indian traditional medicine (Ayurveda) for blood glucose regulation. Most commonly studied Indian herbs for diabetes are *Allium cepa*; *Allium sativum*; *Cajanus cajan*; *Coccinia indica*; *Caesalpinia bonducella*; *Ficus bengalensis*; *Gymnema sylvestre*; *Momordica charantia*; *Ocimum sanctum*; *Pterocarpus marsupium*; *Sweteria chirayita*; *Syzygium cumini*; *Tinospora cordifolia*; *Trigonella foenum-graecum.* Most of them are shown to have anti-diabetic activity (3).

Hypoglycaemic efffects of herbs such as *Momordica charantia*, *Pterocarpus marsupium*, and *Trigonella foenum-greacum* is thought to be through their stimulating or regenerating effects on β-cells or extra-pancreatic effects (4).

Cowplant/Gurmar/Meshasringa/Madhunasini/*Gymnema sylvestre*

Image Courtesy: Mr Vinayaraj V. R.

It is a herb native to India and Sri Lanka. Its Sanskrit name means sweet killer as chewing of the leaves of this plant suppresses the sweet taste. Later, it was found that this property is due to the presence of gymnemic acids.

Studies done in lab animals were encouraging (5,6) and this was further supported by clinical trials in patients.

An extract from *G. sylvestre* was studied in diabetic population (twenty-two patients) on treatment with conventional drug. During the course of 18-22 months of treatment, patients showed significant reduction in blood sugar values, HbA1c, and glycosylated plasma proteins. Most of the patients could also reduce the dosage of conventional medicines and many of them (five patients) were able to discontinue their medications. The increased serum insulin levels of the patients on supplementation suggested β cell repair or regeneration (7).

Gymnema's insulinotropic effect is associated with increased influx of Ca+ into the β cells (8). Studies done with oral gymnema extract (1 g/day for sixty days) induced significant increase in circulating insulin and c-peptide levels. This was also associated with reduction in fasting and postprandial glucose (9,10).

Gooseberry/Amla/*Phyllanthus emblica/Emblica officinalis*

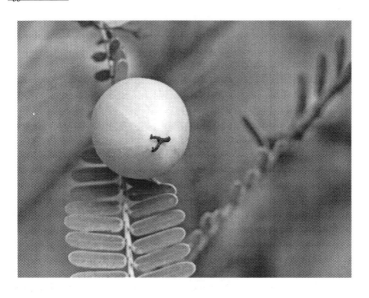

Gooseberry or Amla is widely seen in India and Africa. Its fruit is rich in antioxidants like vitamin C and used traditionally for improving the immunological function and for its anti-aging effects. Available researches show that it is having anti-diabetic and anti-triglyceride properties (11).

Studies done in diabetes-induced rats show that oxidative stress is an important factor in producing diabetic cataracts, and tannin in emblica is able to delay it (12). It is also shown to be having free radical scavenging properties in rat experiments (13).

Black plum/Jamun/Jambul/*Syzygium jambolanum*

Syzygium jambolanum (commonly known as jambol fruit or jamun) is a tree found in India, Pakistan, Southern Asia, and Brazil. Before the invention of insulin, it was used as a main anti-diabetic agent in Europe (14).

Mother tincture of *S. jambolanum* is widely used by Homeopathy practitioners for diabetes management. The seed extract contains glycoside (jamboline), tannin, ellagic acid, and gallic acid as principal ingredients (15).

Experiments with mother tincture of *S. jambolanum* in rats found to be effective in reducing the blood sugar, cholesterol, and as having hepatoprotective effect (16).

Various parts of the fruit were studied for their anti-diabetic effect. Alcoholic extract of the kernel was found to significantly reduce blood sugar, urea, and cholesterol. It was also found to increase glucose tolerance in experimental rats. Whole seed showed moderate hypoglycaemic effect whereas the seed coat did not. The glucose lowering effect was comparable to standard hypoglycaemic drug, glibenclamide (17).

Ivy Gourd/Baby watermelon/*Coccinia indica*

This is a creeper seen widely in India and mentioned in Ayurvedic text books (as Bimba) for diabetic treatment. Double blind studies done with the crushed leaves of *Coccinia indica* showed significant improvement in glycaemic control in diabetic patients (18). Another study done in experimental rats found to have significant effect on plasma vitamin C and reduced glutathion. Its antioxidant and hypoglycaemic effects were found to be greater than glibenclamide.

The mechanism of action of *C. indica* is not well understood, but the herb appears to be insulinomimetic. It has been postulated that the ingredients present in the extract of *C. indica* act like insulin. It achieves control of blood sugar by correcting the elevated enzymes (glucose-6-phosphatase and lactate dehydrogenase) in the glycolytic pathway and by restoring lipoprotein lipase activity in the lipolytic pathway (19).

Yeh et al. assessed the quality of the evidence of *C. indica* for glucose control using American Diabetes Association Criteria for Clinical Guidelines and found to have 'A' rating (supportive evidence with at least one adequate randomised clinical trial) (20).

Studies done with alcoholic extract of *Coccinia cordifolia* (1 g) for ninety days in diabetic people in a double blind placebo controlled randomised trial found that it has hypoglycaemic actions (21).

Bitter Melon/Bitter Gourd/Karela/*Momordica charantia*

Bitter gourd is an edible vegetable mentioned in Ayurveda for diabetes. It has substances like charantin, vincine, polypeptide p, and antioxidants. Hypoglycaemic and metabolic effects of bitter gourd extracts were demonstrated in cell culture, animal, and human studies. Like in many other herbs, its mechanism of action is not clearly understood (22).

Study in diabetic rats with bitter gourd powder mixed in diet (10%) showed a 30% reduction of fasting blood sugar levels (23).

There are structural and functional changes noticed in alimentary tracts of diabetics. This results in increased absorption of intestinal glucose and alteration in brush border disaccharidases. Changes are also noticed in renal cortex of kidneys. The effect of *M. charantia* fruit in activities of renal and intestinal disaccharidases were studied in diabetic-induced rats. Specific activities of intestinal diasccharidases were significantly increased during diabetes and by supplementing bitter gourd in diet showed reduction of certain disaccharidases (maltase, lactase) (24).

Indian Keno Tree/Bijaka/*Pterocarpus marsupium*

Image Courtesy: Mr Vinayaraj V. R.

Pterocarpus marsupium is a medicinal tree belonging to the group called rasayana in ayurvedic system of medicine. Rasayana drugs are immunomodulators and relieve stress in the body. In Ayurveda, aqueous extract of heart-wood of *P. marsupium* is used in the treatment of diabetes.

Compounds like pterostilbene, flavonoids, marsupin, pterosupin, liquiritigenin, epicatechins derived from the plant shows various beneficial effects. Pterostilbene is shown to be hypoglycaemic. Flavonoids are shown to regenerate pancreatic β cells. Marsupin, pterosupin, and liquiritigenin have anti-hyper lipidaemic activity. The active principle epicatechin is shown to enhance insulin release and conversion of pro-insulin to insulin.

In the previous sections we have seen that inflammatory cytokines like TNF-α, IL-1β, and IL-6 are raised in diabetic people. We also saw that activation of innate immunity and chronic systemic inflammation are pathogenic process in type 2 diabetes.

A study done in diabetic rats with aqueous extract of *P. marsupium* significantly reduced hyperglycaemia and TNF-α level. This shows its potential anti-inflammatory and anti-diabetic effects (25).

The beneficial effects of Kino tree in blood glucose, cholesterol, and liver functions are proved through experiments (26).

Cinnamon/*Cinnamomum verum*

Cinnamon is a pungent-smelling spice used commonly in Asian cuisine.

Spices like allspice, cinnamon, bay leaf, cloves, nutmeg, oregano, and black or green tea have been shown to have an insulin-like biological activity. Among these spices, cinnamon has shown to have the highest bioactivity.

Low levels of cinnamon, about 1-6 g/day has got effects of lowering glucose, triglyceride, LDL cholesterol, and total cholesterol levels in subjects with type 2 diabetes.

The bioactive compound in cinnamon responsible for this action is found to be a water-soluble polyphenolic type A polymer (27). This has found to have insulin-like activity and antioxidant effect. Cinnamon has been shown to reduce fasting blood sugar levels, triglycerides, and total and LDL cholesterol levels in patients with type 2 diabetes.

Ranasinghe et al. in a systematic review and meta-analysis of sixteen studies found that cinnamon has potential beneficial effects in reducing postprandial glucose absorption by different mechanisms. It is also found that it lowers body weight, glucose, LDL cholesterol, and HbA1c levels, whereas it increases HDL and circulating insulin levels. It is also shown to be effective against diabetic neuropathy and nephropathy. The authors mention that it does not have significant toxicity in liver and kidneys and has high therapeutic window (28).

Fenugreek/*Trigonella foenum-graeceum*

Fenugreek is extensively used as a spice and herb in Asian countries. Its use is mentioned in herbal systems of Ayurveda and Unani. Fenugreek seeds contain 50% fibre, 30% of it is soluble and 20% insoluble. The fibre slows down the rate of glucose absorption after a meal. Amino acid 4-hydroxyisoleucine extracted from fenugreek seeds is shown to stimulate insulin action (29).

The soluble fibre fraction of fenugreek seed exerts its anti-diabetic action through inhibition of carbohydrate digestion and absorption and improving peripheral utilisation of insulin (30).

When used along with conventional medicines, fenugreek seeds improve glycaemic control and reduce insulin resistance in type 2 diabetic patients. It has also shown to have favourable effect on triglyceride levels (31).

Turmeric/Haridra/*Curcuma longa*

Turmeric is a popular spice in Asian cuisine for ages. It is widely consumed and believed to be having multiple health benefits. Curcumin is the main component in turmeric.

Curcumin is shown to have antioxidant, anti-inflammatory, anti-viral, anti-fungal, antibacterial, and anti-cancer properties. Because of these properties, it has potential role in various diseases like diabetes, allergies, cancers, arthritis, Alzheimer's, and other chronic disorders. It exhibits properties as TNF blocker, vascular endothelial cell growth factor blocker, and human epidermal growth factor receptor blocker. As it is having beneficial effects in many diseases, it can be used as a less expensive, low side-effect alternative to prescription pills (32).

We have already seen the implications of glycation, Maillard reactions, and oxidative stress in pathogenesis of many diseases including diabetes. Alteration in mitochondrial function is seen as a central mechanism in many oxidative stress-related diseases. Damaged mitochondria produce more reactive oxygen species (ROS) and less ATP than normal. It cannot use glucose or lipid and cannot provide cells with ATP. The unutilised glucose, fat, and amino acids accumulate outside the mitochondria, and they undergo more glycation. Antioxidants can slow down these processes and thereby reduce cell damage (33).

Curcuminoids, the main yellow pigments in turmeric, have been used widely in indigenous medicine for the treatment of sprain and inflammation. Curcuminoids are having antioxidant function.

Curcumin and its major metabolite tetrahydrocurcumin (THC) are shown to have protective effects (34). Tetrahydrocurcumin (THC) is shown to have anti-diabetic and anti-hyperlipideamic action in diabetic rats (35). In addition, it could delay development of type 2 diabetes, improve β-cell functions, prevent β-cell death, and reduce insulin resistance in animals.

Studies done in diabetic rats by administering curcumin showed that it significantly reduced blood sugar and HbA1c. It is also effective in reducing oxidative stress by preventing influx of glucose into the polyol pathway, thereby increasing levels of antioxidant enzymes (36). The importance of polyol pathway in diabetes complications is already discussed in previous chapter.

In human studies, curcumin is shown to delay the development of diabetes (37). It is also shown to have glucose lowering effects in overweight or obese diabetic patients. This action is partially due to decrease in serum-free fatty acids, which may result from promoting fatty acid oxidation and utilisation (38).

- Many herbs are found to be having blood sugar lowering function.
- Mechanism of these herbs in the body is complex, so also the effect of diabetes on human body.
- The action of herbs is not restricted to lowering blood sugar.
- There are scientific validations for the efficacy of many herbs.

References

1. Modak M, Dixit P, Londhe J, Ghaskadbi S, Devasagayam TPA. Indian herbs and herbal drugs used for the treatment of diabetes. J Clin Biochem Nutr. 2007;40(3):163-73

2. WHO Expert Committee: Diabetes Mellitus. 2nd rep. Geneva: World Health Organization; 1980 (Tech. Rep. Ser. 646)

3. Grover JK, Yadav S, Vats V. Medicinal plants of India with anti-diabetic potential. J Ethnopharmacol. 2002;81:81-100

4. Saxena A, Vikram NK. Role of selected Indian plants in management of type 2 diabetes: a review. J Altern Complement Med. 2004;10:369-78

5. Okabayashi Y, Tani S, Fujisawa T, Koide M, Hasegawa H, Nakamura T, et al. Effect of *Gymnema sylvestre*, R.Br. on glucose homeostasis in rats. Diabetes Res Clin Pract. 1990;9(2):143-8

6. Shanmugasundaram ER, Gopinath KL, Radha Shanmugasundaram K, Rajendran VM. Possible regeneration of the islets of Langerhans in streptozotocin-diabetic rats given *Gymnema sylvestre* leaf extracts. J Ethnopharmacol. 1990;30(3):265-79

7. Baskaran K, Kizar Ahamath B, Radha Shanmugasundaram K, Shanmugasundaram ER. Anti-diabetic effect of leaf extract from *Gymnema sylvestre* in non-insulin-dependent diabetes mellitus patients. J Ethnopharmacol. 1990;30(3):295-300

8. Liu B, Asare-Anane H, Al-Romaiyan A, Huang G, Amiel SA, Jones PM, Persaud SJ. Characterisation of the insulinotropic activity of an aqueous extract of *Gymnema sylvestre* in mouse beta-cells and human islets of Langerhans. Cell Physiol Biochem. 2009;23(1-3):125-32

9. Al-Romaiyan A, Liu B, Asare-Anane H, Maity CR, Chatterjee SK, Koley N, Biswas T, Chatterji AK, Huang GC, Amiel SA, Persaud SJ,Jones PM. A novel *Gymnema sylvestre* extract stimulates insulin secretion from human islets in vivo and in vitro. Phytother Res. 2010;24(9):1370-6

10. Leach MJ. *Gymnema sylvestre* for diabetes mellitus: a systematic review. J Altern Complement Med. 2007;13(9):977-83

11. Qureshi SA, Asad W, Sultana V. The effect of *Phyllanthus emblica* Linn on type-II diabetes: triglycerides and liver-specific enzyme. Pakistan J Nutr. 2009;8(2):125-8

12. Suryanarayana P, Saraswat M, Petrash JM, Reddy GB. Emblica officinalis and its enriched tannoids delay streptozotocin-induced diabetic cataract in rats. Mol Vis. 2007;13:1291-7

13. Nain P, Saini V, Sharma S, Nain J. Anti-diabetic and antioxidant potential of emblica officinalis Gaertn. leaves extract in streptozotocin-induced type 2 diabetes mellitus (T2DM) rats. J Ethnopharmacol. 2012;142(1):65-71

14. Baliga MS, Fernandes S, Thilakchand KR, D'souza P, Rao S. Scientific validation of the anti-diabetic effects of *Syzygium jambolanum* DC (black plum), a traditional medicinal plant of India. J Altern Complement Med. 2013;19(3):191-7

15. Shanbhag DA, Khandagale AN. Application of HPTLC in the standardization of a homoeopathic mother tincture of *Syzygium jambolanum*. J Chem Pharm Res. 2011;3:395-401

16. Maiti S, Ali KM, Jana K, Chatterjee K, De D, Ghosh D. Ameliorating effect of mother tincture of *Syzygium jambolanum* on carbohydrate and lipid metabolic disorders in streptozotocin-induced diabetic rat: homeopathic remedy. J Nat Sci Biol Med. 2013;4(1):68-73

17. Ravi K, Sivagnanam K, Subramanian S. Anti-diabetic activity of *Eugenia jambolana* seed kernels on streptozotocin-induced diabetic rats. J Med Food. 2004;7(2):187-91

18. Azad Khan AK, Akhtar S, Mahtab H. Treatment of diabetes mellitus with *Coccinia indica*. Br Med J. 1980;260:1044

19. Venkateswaran S, Pari L. Effect of *Coccinia indica* leaf extract on plasma antioxidants in streptozotocin-induced experimental diabetes in rats. Phytother Res. 2003;17(6):605-8

20. Yeh GY, Eisenberg DM, Kaptchuk TJ, Phillips RS. Systematic review of herbs and dietary supplements for glycemic control in diabetes. Diabetes Care. 2003;26:1277-93

21. Kuriyan R, Rajendran R, Bantwal G, Kurpad AV. Effect of supplementation of *Coccinia cordifolia* extract on newly detected diabetic patients. Diabetes Care. 2008;31(2):216-20

22. Krawinkel MB, Keding GB. Bitter gourd (*Momordica charantia*): a dietary approach to hyperglycemia. Nutr Rev. 2006;64(7 Pt 1):331-7

23. Shetty AK, Kumar GS, Sambaiah K, Salimath PV. Effect of bitter gourd (*Momordica charantia*) on glycaemic status in streptozotocin induced diabetic rats. Plant Foods Hum Nutr. 2005;60(3):109-12

24. Kumar Shetty A, Suresh Kumar G, Veerayya Salimath P. Bitter gourd (*Momordica charantia*) modulates activities of intestinal and renal disaccharidases in streptozotocin-induced diabetic rats. Mol Nutr Food Res. 2005;49(8):791-6

25. Halagappa K, Girish HN, Srinivasan BP. The study of aqueous extract of *Pterocarpus marsupium* Roxb. on cytokine TNF-α in type 2 diabetic rats. Indian J Pharmacol. 2010;42(6):392-6

26. Dhanabal SP, Kokate CK, Ramanathan M, Kumar EP, Suresh B. Hypoglycaemic activity of *Pterocarpus marsupium* Roxb. Phytother Res. 2006;20(1):4-8

27. Anderson RA, Broadhurst CL, Polansky MM, Schmidt WF, Khan A, Flanagan VP, et al. Isolation and characterization of polyphenol type-A polymers from cinnamon with insulin-like biological activity. J Agric Food Chem. 2004;52:65-70

28. Ranasinghe P, Jayawardana R, Galappaththy P, Constantine GR, de Vas Gunawardana N, Katulanda P. Efficacy and safety of 'true' cinnamon (*Cinnamomum zeylanicum*) as a pharmaceutical agent in diabetes: a systematic review and meta-analysis. Diabet Med. 2012;29(12):1480-92

29. Baquer NZ, Kumar P, Taha A, Kale RK, Cowsik SM, McLean P. Metabolic and molecular action of *Trigonella foenum-graecum* (fenugreek) and trace metals in experimental diabetic tissues. J Biosci. 2011;36(2):383-96

30. Hannan JM, Ali L, Rokeya B, Khaleque J, Akhter M, Flatt PR, et al. Soluble dietary fibre fraction of *Trigonella foenum-graecum* (fenugreek) seed improves glucose homeostasis in animal models of type 1 and type 2 diabetes by delaying carbohydrate digestion and absorption, and enhancing insulin action. Br J Nutr. 2007;97(3):514-21

31. Gupta A, Gupta R, Lal B. Effect of *Trigonella foenum-graecum* (fenugreek) seeds on glycaemic control and insulin resistance in type 2 diabetes mellitus: a double blind placebo controlled study. J Assoc Physicians India. 2001;49:1057-61

32. Aggarwal BB, Sundaram C, Malani N, Ichikawa H. Curcumin: the Indian solid gold. Adv Exp Med Biol. 2007;595:1-75

33. Edeas M, Attaf D, Mailfert AS, Nasu M, Joubet R. Maillard reaction, mitochondria and oxidative stress: potential role of antioxidants. Pathol Biol (Paris). 2010;58(3):220-5

34. Osawa T, Kato Y. Protective role of antioxidative food factors in oxidative stress caused by hyperglycemia. Ann N Y Acad Sci. 2005;1043:440-51

35. Pari L, Murugan P. Antihyperlipidemic effect of curcumin and tetrahydrocurcumin in experimental type 2 diabetic rats. Ren Fail. 2007;29(7):881-9

36. Arun N, Nalini N. Efficacy of turmeric on blood sugar and polyol pathway in diabetic albino rats. Plant Foods Hum Nutr. 2002 Winter;57(1):41-52

37. Chuengsamarn S, Rattanamongkolgul S, Luechapudiporn R, Phisalaphong C, Jirawatnotai S. Curcumin extract for prevention of type 2 diabetes. Diabetes Care. 2012;35(11):2121-7

38. Na LX, Li Y, Pan HZ, Zhou XL, Sun DJ, Meng M, et al. Curcuminoids exert glucose-lowering effect in type 2 diabetes by decreasing serum free fatty acids: a double-blind, placebo-controlled trial. Mol Nutr Food Res. 2013;57(9):1569-77

CHAPTER 10

Sweeteners

Fondness to sweet is inborn to humans, and it dates back to the early civilisations. Mankind used honey as a sweetener in ancient cultures like Greek, Indian, and Chinese. Later it was replaced by sucrose or common sugar, which was obtained from sugar cane. But it is also sourced from sugar beet, a practice started during world war. The scarcity of common sugar during world wars prompted us to look into alternate sources of sweeteners. This paved the way for artificial sweeteners in the food industry.

The artificial sweeteners can be broadly divided into two categories: *first generation* and *second generation*.

Saccharine, cyclamate, and aspartame are first generation and acesulfame-K, sucralose, alitame, and neotame are second generation.

Saccharine

Saccharine is the first artificial sweetener synthesised in 1879 by Remsen and Fahlberg (1). This coal-tar derivative (benzoic sulfomide) was found out accidentally when

Fahlberg was looking for the oxidative mechanism of tolunesulfonamide. Saccharine is 300 times sweeter than the common sugar. It was widely used until 1907 in canned foods but ultimately banned from its use as a food additive. During World War I, it was given approval as there was shortage of cane sugar. In early 1950s its usage became more widespread as we became more 'calorie conscious'. With this came a new section of food products labelled as diet/low-calorie products. Companies jumped into this bandwagon because of two main reasons: one, it was fashionable at that time (still it is fashionable!) and second, it was cheap compared to sugar. From 1963 to 1967, artificially sweetened soft drinks tripled their market share. It is used to sweeten various products, including soft drinks, baked goods, jams, chewing gum, canned fruit, candy, dessert toppings, and salad dressings. Saccharin is also used in cosmetic products (e.g., toothpaste, mouthwash, and lip gloss), vitamins, and medications. Saccharine has a roller-coaster history. It had its peaks and dips, was very popular, and at times banned by regulatory authorities. It is still banned in Canada.

According to me, any *edible* product which is made in the lab would not be safe and nutritious as a natural *edible* product. There were reports and studies showing bladder cancer among laboratory animals tested with saccharin, which later were refuted by some researchers. The acute (immediate) toxicity observed after using saccharine are nausea, vomiting, and diarrhoea. The long-term (chronic) toxicities noticed are cancer in offspring of breast-fed and low birth-weight babies, bladder cancer, and hepatotoxicity.

Cyclamate

The bitter aftertaste of saccharine prompted researchers to look for an alternative. A breakthrough was achieved with recognition of cyclamate in 1950s, which provided better taste than saccharin.

A product named Sweet'N Low—which is a combination of saccharine and cyclamate or cyclamate alone (in Canada where saccharine is banned) mixed with some additives—became a huge success in the United States.

But in 1970, FDA banned cyclamate in all food products, as it was reported to have some cancer risk in laboratory animals. Cyclamate is still used in other countries.

Aspartame

Aspartame was approved in 1981, and it is marketed as 'Nutrasweet'.

It is 180 times sweeter than normal sugar (sucrose). Aspartame was discovered accidentally by James Schlatter in 1965 while testing an anti-ulcer drug for G. D. Searle & Company. It is a synthetic molecule composed of two amino acids: aspartic acid (40%), and phenylalanine (50%). The rest of it (10%) is methanol. Its chemical name is L-aspartyl-L-phenylanyl-methyl ester.

It took nearly fifteen years for FDA to approve it as a sweetener. It got clearance from FDA in 1981 and seems that there were lot of controversies and politics behind its approval (2,3).

The widely quoted studies (Leone et al.(4); Schiffman et al.(5)) supporting the aspartame was criticised because of its methodical flaws and grants from the NutraSweet company (2).

We cannot brush aside the studies and articles raising concerns of its safety and effects in human brain. It has been shown in experiments that aspartame can significantly raise phenylalanine levels in rat's brain (3).

The acceptable daily intake (ADI)—the safe quantity an individual can consume—is currently set as 50 mg/kg body weight. Aspartame is registered as a food additive with FDA. As per FDA rules, the manufacturers are not obliged to monitor adverse reactions, submit reports to FDA about adverse reactions, or to conduct research if the product is an additive. To qualify as a food additive, the substance should be physiologically inert. But unfortunately, aspartame is not an inert substance!

Substances are studied in rodents to find out their side effects. But what if the metabolic pathways are different in species studied compared with humans? There is difference in speed of metabolising the amino acid phenylalanine to tyrosine in humans and rodents. In humans, aspartame raises blood and brain phenylalanine concentration, but in rodents, it raises tyrosine, which can neutralise the effects of phenylalanine in brain. So there is no logic in studying the safety profile of aspartame in rats for application in humans. Plasma phenylalanine levels are not influenced by the regulatory mechanisms in the body. So phenylalanine measured in plasma reflects the amount absorbed from recently consumed food (3). Even though phenylalanine is a protein constituent, taking a protein meal would not raise its concentration in brain. But consuming aspartame in ADI levels would increase brain phenylalanine by three folds (Stegnic et al)! When a beverage containing aspartame is consumed along with a high-carbohydrate, low-protein dessert—a common combination, the amount of phenylalanine in the brain is doubled (Yokogoshi et al)!

Phenylalanine in the brain reduces the synthesis of certain neurotransmitters by inhibiting key enzymes like tyrosine hydroxylase. High circulating levels of phenylalanine in the brain reduce production of brain catecholamines and serotonins which are important in mood and controlling seizure. Dysfunction in catecholamine neurotransmission are important in some neurologic and neuropsychiatric disorders.

There are many ways in which aspartame produces its brain effects. These are by disturbing amino-acid metabolism, protein structure and metabolism, integrity of nucleic acids, neuronal function, endocrine balances, and by changing the brain concentrations of catecholamines. It is also seen that aspartame and its breakdown products cause nerves to fire excessively, which indirectly causes seizures (6).

Common side effects mentioned for aspartame are headaches, change in mood, nausea, cramps and abdominal pain, diarrhoea, fatigue, and seizures. There are plenty of books written on this topic. One such example is *Aspartame Disease: An ignored epidemic* by H. J. Roberts.

Acesulfame-K

Acesulfame potassium was discovered in 1967 by a chemist named Karl Clauss. It was approved by FDA as table-top sweetener in 1988, in beverages by 1998, and as a general sweetener in foods by 2006. It is 200 times sweeter than sugar and has zero calories.

In toxicity studies in mice (Mukherjee and Chakrabarti (7)), it did not show significant chromosomal changes in dosage of 15 mg/kg body weight. But in higher dosages, it is found to be toxic to genes and cause chromosomal changes.

It is now being added in around 6,000 products, including candies, baked goods, frozen desserts, beverages, and cough drops. Popular brands are Sweet One and Sunett.

Sucralose

Sucralose was discovered in 1976 by a British sugar company Tate and Lyle. It is 600 times sweeter than sugar and contains no calories. It was approved by FDA in 1998 to be used in fifteen categories, including table-top sweetener. It is marketed under the brand name Splenda.

The sucralose is zero calories, but the product made out of it has got fillers like maltodextrin and glucose apart from sucralose (1.1%). The fillers are required as the quantity of sweetener put in a sachet (may be a few grains) is very small because of its sweetness. With the addition of fillers, each packet makes about having four calories.

Sucralose is used in around 4,000 products like beverages, baked goods, breakfast cereal, tea, coffee, chewing gums, desserts, and even in pharmaceutical products.

In a study conducted on a group of rats, researchers found that Splenda suppresses intestinal microflora of these animals. Splenda increases expression of intestinal chemical transporter P-glycoprotein (P-gp) and CYP 450 isoenzymes (CYP 3A4 and CYP 2D1). They also noticed a reduction of intestinal pH. The P-gp is responsible for multidrug resistance in chronic anti-cancer drug treatment. The CYP enzymes are involved in metabolism of various drugs. These indicate there is a potential for Splenda to interfere with the bioavailability of drugs and nutrients. All these effects were noticed in a dosage equivalent to the approved dosage of sucralose by the FDA (8). The reduction of beneficial bacteria in the gut can cause immunological and intestinal effects. In a review article published in the *World Journal of Gastroenterology*, the author assumes the unrestricted use of sucralose as one of the explanations for the recent increase of irritable bowel disease observed in children (9).

Neotame

It is a newest artificial sweetener, a derivative of aspartame. It is 7,000 to 13,000 times sweeter than sugar. FDA gave approval for its usage as general purpose sweetener in 2002. Unlike the parent compound aspartame, it reduces the availability of phenylalanine in tissues.

In a review article titled 'Artificial sweeteners—do they bear a carcinogenic risk?' M. R. Weihrauch and V. Diehl mention that heavy artificial sweetener usage of >1,680 mg/day leads to an increase of bladder cancer in humans (relative risk of 1.3). They say it is difficult to precisely determine the culprit as many artificial agents are combined in current foods. The recent approval of many such substances makes it impossible to establish any epidemiological evidence about a possible cancer risk. It is too early to establish a link between artificial sweeteners and cancers as we started using them very recently (10).

Natural Sweetener

Stevia

It is extracted from the Stevia plant *Stevia rebaudiana,* which is native to South America. It was used by Guarani tribe of Paraguay as a sweetener in their drinks for centuries. They named it Ka'a he'e (means sweet herb). It has been traditionally used in Paraguay and Brazil as a sweetener for tea (11).

It was first researched by a Spanish botanist and physician Petrus Jacobus Stevus. The Latin name Stevia was originated from his surname. Japanese started cultivating it in early 1970, and first commercial sweetener from Stevia produced in as early as 1971, and since then, they were using it extensively in their food as a sweetener. Japanese consume more Stevia than any other country, and Stevia accounts for 40% of their sweetener market.

Stevia's medicinal uses include regulation of blood sugar, prevention of hypertension, treatment of skin disorders, and prevention of tooth decay. It is also having antibacterial and antiviral properties (12). A study conducted in healthy volunteers show that it does not have anti-hypertensive effects (13). Stevia was found to have antioxidant, anti-inflammatory, immunomodulatory, bactericidal, and anti-hypertensive properties in laboratory experiments.

About one-fourth teaspoon of ground leaf is equivalent to one teaspoon of sugar. Compared to sugar, it does not produce tooth decay, and it is a non-calorie sweetener. The plant gets its sweetness from alkaloids stevoside, rebaudiosides (A, B, C, D, and E), dulcoside, and steviolbioside. Stevoside has a bitter aftertaste and it is 250-300 times sweeter than sugar. But rebaudioside doesn't have bitter aftertaste (14).

Stevia reduces blood sugar levels by enhancing insulin secretion and increasing utilisation of insulin in peripheral tissues and muscles.

- Plenty of sweeteners are available in the market.
- Most of them are synthetic preparations.
- There are controversies regarding their safety profile.
- Natural sweeteners from the plants are having other beneficial effects in diabetes.

References

1. http://en.wikipedia.org/wiki/Saccharin
2. Walton, Ralph G., Robert Hudak, Ruth Green-Waite. Adverse reactions to aspartame: double-blind challenge in patients from a vulnerable population. Biol Psychiatry. 1993;34:13-7
3. Humphries P, Pretorius E, Naude H. Direct and indirect cellular effects of aspartame on the brain. Eur J Clin Nutr. 2008;62:451-62
4. Leon AS, Hunninghake DB, Bell C, Rassin DK and Tephly TR. Safety of long-term large doses of aspartame. Arch Int Med. 1989;149(10): 2318-24.
5. Schiffman SS, Buckley CE 3rd, Sampson HA, Massey EW, Baraniuk JN, Follett JV, Warwick ZS. Aspartame and susceptibility to headache. N Engl J Med. 1987;317(19):1181-5.

6. Maher TJ, Wurtman RJ. Possible neurologic effects of aspartame, a widely used food additive. Environ Health Perspect. 1987;75:53-7

7. Mukherjee A, Chakrabarti J. In vivo cytogenetic studies on mice exposed to acesulfame-K—a non-nutritive sweetener. Food Chem Toxicol. 1997;35(12):1177-9

8. Abou-Donia MB, El-Masry EM , Abdel-Rahman AA, McLendon RE, Schiffman SS. Splenda alters gut microflora and increases intestinal P-glycoprotein and cytochrome P-450 in male rats. J Toxicol Environ Health A. 2008;71:1415-29

9. Qin X. Etiology of inflammatory bowel disease: a unified hypothesis. World J Gastroenterol. 2012;18(15):1708-22

10. Weihrauch MR, Diehl V. Artificial sweeteners— do they bear a carcinogenic risk? Ann Oncol. 2004;15:1460-65

11. http://en.wikipedia.org/wiki/Stevia

12. Kujur RS, Singh V, Ram M, Yadava HN, Singh KK, Kumari S, Roy BK. Anti-diabetic activity and phytochemical screening of crude extract of Stevia rebaudiana in alloxan-induced diabetic rats. Pharmacognosy Res 2010;2:258-63

13. Geuns JM, Buyse J, Vankeirsbilck A, et al. Metabolism of stevioside by healthy subjects. Exp Biol Med (Maywood). 2007;232(1):164-73

14. Madan S, Ahmed S, Singh GN, Kohli K, Kumar Y, Singh R, Garg M. Stevia rebaudiana Bertoni—a review. Indian J Nat Prod Resour. 2010;1(3):267-86

CHAPTER 11

Conclusion

During our sojourn through this book, we started with the startling statistics of the epidemic 'diabesity', its vast implications in many metabolic derangements in the system, and how our diet and lifestyle can influence and modify it. It is clear that these clusters of disorders are not the result of 'one cause one effect' theory. It is perpetuated by multitude of factors starting from susceptible genes, modified and influence significantly through nutrients in diet, the exercises one carries out, and the environment one is exposed to.

We have also seen that there are favourable changes in epigenome with diet and physical activity. This is a strong argument for considering them seriously rather than mere means of losing some calories and shedding the extra fat.

Medical practitioners have to stress more about the importance of lifestyle and dietary advice before picking the pen to write a pharmaceutical prescription.

The current approach to search for the 'magic bullet' and driving this message into the minds of people might be beneficial for some companies. Most of the companies are happy if millions of patients pin their hopes on a pill, believing it would cure their metabolic maladies. I am not discounting the role of these prescription drugs which serve as life-saver in many patients.

Let us all hope that practitioners and patients become more logical when they consider options for managing blood sugar.

Ten worst things to be avoided by diabetics

1. Irregular eating habits
2. Being sedentary
3. Stress
4. Inadequate sleep
5. Foods with high GI and GL and foods with high fructose (like high-fructose corn syrup)
6. Foods which are high in AGEs: deep fried, foods having brown crust and laden with trans-fats
7. High temperature dry cooking
8. Overindulging in sweets and trying to control it with extra shot of medications
9. Overloading the system with toxins, medications, cocktail of supplements
10. Very low blood sugar levels

Ten best things to be followed by diabetics

1. Eat regularly and never skip a meal. Try to make your carbs more complex by adding more vegetables to them

2. Control portion size. Small portion every 2-3 hours is good for people on insulin
3. Maintain proper hydration by drinking plain water—at least 1.5-2 litres per day
4. Do regular good quality exercises, preferably with safe sun exposure.
5. Sleep at least seven hours a day
6. Eat more fresh and fibrous foods than processed, preserved, and refined foods
7. Slow moist cooking rather than express cooking in high temperature—Steaming and making curries rather than frying, grilling
8. Natural fats acceptable provided you don't modify them
9. Maintain very good HbA1c levels and average sugar values
10. Consume supplements like omega-3 fatty acids, Vit D, Cr, Mg, Zn as per advice of your physician who understands their importance

Best foods with low GI and GL

-   Whole grain foods than refined flour
-   Whole fruits than fruit juices
-   Green leafy vegetable
-   High-protein foods: fish, lentils, beans, eggs
-   Crunchy vegetable than starchy vegetables
-   Fishes which are rich in omega-3 fats: sardine, salmon, mackerel, tuna
-   Healthy fats: nuts, seeds, fish with omega-3 fats, non-processed fats

- Beans, lentils, and pulses which are normally rich in protein and minerals
- Spices: normally rich in anti-oxidants with high ORAC values and low calorie